ELECTROMAGNETICS THROUGH THE FINITE ELEMENT METHOD

A Simplified Approach Using Maxwell's Equations

ELECTROMAGNETICS THROUGH THE FINITE ELEMENT METHOD

A Simplified Approach Using Maxwell's Equations

José Roberto Cardoso

ESCOLA POLITECNICA DA UNIVERSIDADE DE SAO PAULO

CRC Press
Taylor & Francis Group
Boca Raton London New York

CRC Press is an imprint of the
Taylor & Francis Group, an **informa** business

CRC Press
Taylor & Francis Group
6000 Broken Sound Parkway NW, Suite 300
Boca Raton, FL 33487-2742

First issued in paperback 2019

© 2017 by Taylor & Francis Group, LLC
CRC Press is an imprint of Taylor & Francis Group, an Informa business

No claim to original U.S. Government works

ISBN-13: 978-1-4987-8357-6 (hbk)
ISBN-13: 978-0-367-89016-2 (pbk)

Library of Congress Cataloging-in-Publication Data

Names: Cardoso, José Roberto.
Title: Electromagnetics through the finite element method : a simplified approach using Maxwell's equations / José Roberto Cardoso.
Description: Boca Raton : Taylor & Francis, 2016. | "A CRC title." | Includes bibliographical references and index.
Identifiers: LCCN 2016020302 | ISBN 9781498783576 (alk. paper)
Subjects: LCSH: Electromagnetism--Mathematics. | Maxwell equations--Numerical solutions. | Finite element method.
Classification: LCC QC760 .C358 2016 | DDC 537.01/51825--dc23
LC record available at https://lccn.loc.gov/2016020302

Visit the Taylor & Francis Web site at
http://www.taylorandfrancis.com

and the CRC Press Web site at
http://www.crcpress.com

To Jandira T. M. Cardoso, my sweet wife and inspiration

Contents

Preface

I remember the time when Prof. Jean-Claude Sabonnadiere from the LEG (Laboratoire d'Electrotechnique de Grenoble/France) was in a conference at the Polytechnique School of the University of São Paulo. I was a PhD candidate and my thesis subject was the application of the finite element method (FEM) on the design of electromagnetic devices. Prof. Sabbonadiere was an icon in numerical methods for electromagnetism. Few researchers in Brazil were working on this in 1982, when he visited our university. After his speech, I had a private meeting with him for discussing some aspects of FEM those I had doubts about. To talk privately with one of the first engineers who developed the first application of FEM in electromagnetism was a memorable event. Prof. Sabonnadiere and M.V. Chari independently published the first two papers in the *IEEE Transaction on Power Apparatus and Systems* about the application of 2D FEM in electromagnetism.

These papers triggered an explosion of research in the computation of electromagnetic fields in the early 1970s. Researches around the world created the first specialized labs in this field. Groups from France, the United Kingdom, Germany, the United States, Canada, and Japan rapidly led this growing research line. In Brazil, by the middle of 1980s, some groups emerged in the states of São Paulo, Santa Catarina, and Minas Gerais, whose leaders came back to the country after completing their PhD programs abroad. In São Paulo, with the support of collaborators, I created the LMAG—Applied Electromagnetic Laboratory in 1987 after my return from a professional traineeship at LEG where I worked with Prof. Sabonnadiere, J.L. Coulomb, and G. Meunier.

When you try to create a research lab, you must get students first. When you are not a famous researcher, in order to attract them you have to teach fast enough the rudiments of the research they will face, giving the students enough content to enable them to work independently. At that time, the existent formalisms of FEM for electromagnetics were the variational technique (which searches an extreme of a functional energy that is not easy to obtain) and the weighted residual method (which applies a mathematical development without any physical meaning). I confess it was too difficult to attract the first engineering students to the task.

Many students of electrical engineering consider electromagnetism a nonattractive discipline because they do not see immediate application importunities in this field, but also they find the mathematical formalism to be difficult. We observe the same problem involving numerical methods for electromagnetism. We hope the reader changes this when they reach the final words of this book.

If you read *The Feynman Lectures on Physics*, you will agree with me. The thought that all new scientific knowledge is difficult to introduce is not true; there is always a simple way of presenting it. This stimulated me to come up with an easy approach to introducing the FEM application in electromagnetism in order to help students understand the beauty of electromagnetic theory.

I was enlightened when I read the paper of Kao et al. [3], who described the FEM formulation for evaluating the steady-state temperature distribution in a body using only the differential equation that governs the phenomenon; no variational minimization or weighted residual method was used, only the heat differential equation involving the temperature was used. I thought that this brilliant idea could be applied to electromagnetic equations. In a few days, I developed the 2D FEM formulation for static state in time for publishing the magnetostatic case as an appendix in the final document of my PhD thesis. After that, the other formulations were developed and a small book was printed.

This methodology became very popular not only in our engineering school but also in other Brazilian universities and even among engineers who were working for electrical devices manufacturers. It is after almost 30 years of teaching FEM for electromagnetics based on this formalism for both undergraduate and graduate students this book is being published. I hope it helps your learning of this fascinating numerical method for electromagnetism. This kind of knowledge is not only useful in designing electrical devices but also in advanced technological simulations in research laboratories around the world.

Acknowledgments

I thank all researchers and students who contributed directly or indirectly with suggestions to the development of the FEM methodology presented here. We had very fruitful discussions on the presentation of FEM analysis with our young students in the Polytechnique School at the University of São Paulo. To mention all of them is not possible because this work started in 1986(!) and almost a hundred researchers and students have worked with me since. I am deeply grateful for their help in increasing the knowledge about this exciting kind of research, which led to the establishment of the LMAG—Applied Electromagnetic Laboratory, one of the most important electromagnetic laboratories in Brazil.

I also thank Ashley Gasque, the acquisitions editor of Electrical Engineering & Optical Sciences, for introducing this book to the editorial board at CRC Press/ Taylor & Francis Group.

1 Steps for Finite Element Method

1.1 INTRODUCTION

In 1956, the necessity to study the stiffness and deformations of aircrafts' fuselage for the first time led to the use of the finite element method (FEM), which is credited to John Argyris (1913–2004) (www.argyrisfoundation.org). This method was diffused by Turner and Clough [1], and it is currently one of the most powerful tools used for both the conception and analysis of engineering problems.

Regardless of its conception by mechanical engineers, FEM's large-scale diffusion occurred in the field of civil engineering, in which it was used, for example, for the analysis of huge concrete structures, such as dams, tunnels, and bridges. The first book on FEM was written by Olgierd Cecil Zienkiewicz (1921–2009) [2]. This book provided an academic approach to the method, which was established as a strict mathematical formula, besides being applied in other engineering fields, such as thermal studies in mechanical engineering and percolation on flow in inhomogeneous media. In electrical engineering, the first FEM application was presented by Peter Peet Sylvester (1935–1996) in an article published in 1969 in the *IEEE Transaction on Power System* [3].

Even though Zienkiewicz is credited with editing the first book on FEM, he also had the responsibility of presenting it through a complex and dry mathematical formulation, not adequate for the understanding of engineering students. This led to the method being discussed only in graduation courses and in academic environments, such as research labs. The situation in electrical engineering was not that different: Sylvester's text presented FEM applied to electromagnetism based on the use of the basic concepts of variational calculus, one of the formulations presented by Zienkiewicz, which was also not accessible to undergraduate students. The approaches developed by Sylvester, and others after him, in discussing the application of FEM to electromagnetic studies always introducing them as sophisticated mathematical methodologies created an impression that FEM for electromagnetics was a difficult subject and reserved only to those gifted with advanced mathematics knowledge, when the reality was completely different.

Any new knowledge is not deep enough to search for a simplified alternative, even if that alternative is limited in the beginning. This leads to the need for further research and intense diffusion, which is the beauty of this science. Advanced studies on the use of FEM in three-dimensional electromagnetic fields in the 1990s paved the way for even more advanced mathematical techniques, which, regardless of their elegance, jettisoned undergraduate students from the understanding of this methodology.

Over time, new techniques were introduced to simulate some of the most important phenomena present in electrical engineering. The simplest cases are the static electromagnetic fields, which are not time dependent and are characterized by electrostatics, magnetostatics, and electrokinetics (flow of DC current) studies. Electromagnetic phenomena in a quasi-static regime are the ones present in most industrial settings, mainly those related to designing static and rotating electric machines as well as associated devices.

Eventually, electromagnetic phenomena in a time-dependent state like wave guides and resonant cavities are important for the study of communication devices and antennas and also for solving interference and electromagnetic compatibility problems. In Chapter 2, we discuss the basic equations, based on Maxwell equations, that govern each of these studies, and which will support our further development of discussion in this book.

The main objective of the author is to introduce and familiarize this numeric method to a growing number of interested people. For this the author has dedicated himself, for the last three decades, toward the search of an appropriate way of introducing this method for electrical engineering undergraduate students, taking into account the limitations of mathematical skills of these students, and helping first-time learners of this method.

Our goal is to develope the FEM for solving dedicated electromagnetic problems for the undergraduate electric engineering student based on the basic laws of electromagnetism without using sophisticated mathematical resources. Significant advances for the FEM occurred when high-performance workstations appeared in the 1980s. These workstations had high-performance graphics computing and made the analysis of the problem solving easier using images. As a result, reduction in design time, more reliable solutions, reduction in the number of prototypes, and integration with other productive systems happened. Thus, the use of FEM is closely connected to computing resources that made possible, with impressive speed, the resolution of large, linear systems equations, which made feasible the analysis of more number of design alternatives in short time with accuracies never achieved before. Finally, this book gathers and consolidates data generated from the LMAG-Applied Electromagnetic Laboratory of Escola Politecnica da Universidade de Sao Paulo (Polytechnic School of the University of Sao Paulo) [8,12]. It also brings, essentially, the author's experience of FEM teaching to his students from both undergraduate and graduate programs in electrical engineering. Moreover, the popularity of this methodology can be seen by the increasing number of enthusiastic young people taking up graduate studies for the development of advance methodologies of FEM application involving modern techniques of optimization and multiphysical couplings in two or three dimensions.

1.2   STEPS FOR FINITE ELEMENT METHOD

The FEM requires, in general, the accomplishment of a sequence of steps, which if duly concluded leads to an accurate solution of the problem. These steps guide the reader searching a solution for an electromagnetic problem, and they may have, in some cases, specific nuances for different computational tools [26]. Nevertheless, we present these steps as appropriate to studies of electromagnetic phenomena. They can also, with convenient associations, be adapted to solving other problems, for example, those related to structural analysis in civil engineering and thermal problems in mechanical engineering.

1.2.1 STEP 1: 2D VERSUS 3D

In the real world, all electromagnetic structures are three dimensional; in other words, the fields produced by sources present on the device are electromagnetic fields (compact form of describing, generically, the simultaneous presence of both or singular presence of the electric and magnetic fields) on three directions of a spatial coordinate system. However, what happens is that more than 90% of electric devices present a field distribution with a strong preferential trend in only two directions; for this situation the analysis of the field distribution in a single section of the device is enough to provide, with accuracy, all the information required for the forecast of its performance. Thus, the decision to choose 2D or 3D rests with the designer, as professional experience with the design is decisive on these occasions. As criteria for the orientation, when the field distribution is the same in parallel cross section on the bigger portion of the active part of the equipment, the selection of the 2D representation is justified and recommended. These types of situations occur frequently in rotating electric machines, which present plane symmetry; in other words, the magnetic field distribution is repeated in planes parallel to the cross section. A similar situation also occurs in devices with a cylindrical geometry, for example, solenoids, in which the field distribution repeats in planes passing by their longitudinal axes. In this case, the device presents an axisymmetric situation. When it is not possible to identify any of the symmetries indicated, the three-dimensional solution must be considered, and the entire domain will be considered for the study of the phenomenon. Devices with small dimensions are normally treated like problems with a three-dimensional symmetry because in these cases there is no spatial direction preference for the distribution of electromagnetic fields. Figure 1.1 shows an example of each case to highlight differences.

1.2.2 STEP 2: DEFINITION OF THE DOMAIN

Electromagnetic devices, through their voltage or current sources, are submitted to electric or magnetic fields that can be totally confined to the interior of the devices, for rotating electric machines, or extend to infinite, for fields produced by aerial transmission lines. In the former, the study domain is a closed domain, and in the latter, the study domain is an unbounded domain (Figure 1.2). The FEM is a methodology that requires a closed domain surrounded by a closed line for a two-dimensional case or a closed surface for a three-dimensional case, so that its definition in cases of confined fields is evident; in other words, the device's own geometry is the domain of the study. For problems where the fields extend to infinite, we need to delimit the domain of the study by applying a closed line, separated sufficiently from the device, from which we can consider the field small enough outside the domain.

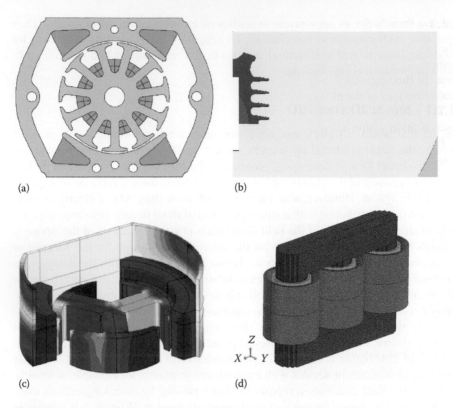

FIGURE 1.1 (a) Plane—direct current motor; (b) axisymmetric—glass insulator from transmission line; and (c, d) three dimension—permanent magnet micromotor and three-phase transformer.

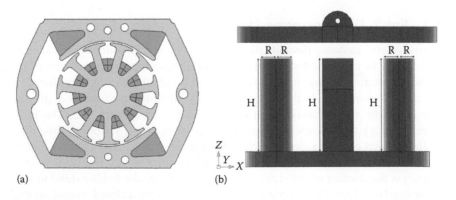

FIGURE 1.2 (a) Closed domain—direct current motor and (b) unbounded domain—electromagnetic actuator.

1.2.3 STEP 3: SELECTION OF THE TYPE OF ELEMENT AND DISCRETIZATION

The following step is the subdivision of the domain into small subdomains (the finite element), a process known as "discretization" performed using well-established criteria. In this step, the designer's experience is explored. It is important to have a qualitative idea of the problem because it will accelerate the search for the solution.

The process will be started by choosing the type of the element to be used for the discretization within a reduced set of few types of finite element geometries.

For two-dimensional problems, the simplest element is the triangle. It is also possible to use the curve-sided triangle (Figure 1.3), the four-sided polygon and curved four-sided polygon. The simultaneous use of different types of finite elements is possible, but one must take into account the compatibility between the common sides of elements.

As a rule, although there are exceptions that will not be discussed here, a two-dimensional domain can be discretized simultaneously using the elements of straight sides, for example, a triangle and a four-sided polygon, or using both elements. The simultaneous use of straight and curve-sided elements presents compatibility only for particular situations; therefore, this should be avoided. For three-dimensional problems (Figure 1.4), the simpler finite element is a tetrahedron, which is presented with two types: a straight tetrahedron or a curved tetrahedron, or a hexahedral element can also be used, which also presents curved sides. The general rule, applied for the two-dimensional case, of avoiding the simultaneous use of different kinds of finite elements for the discretization of the continuous domain must be applied for the three-dimensional case too.

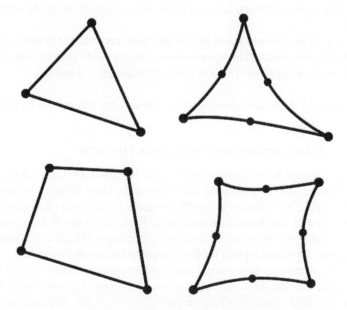

FIGURE 1.3 Types of finite elements for two-dimensional problems.

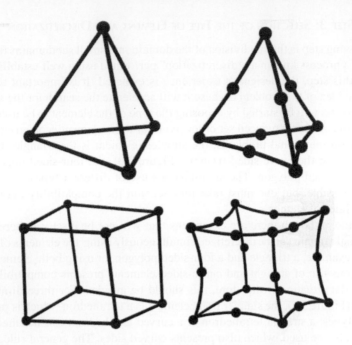

FIGURE 1.4   Types of finite elements for three-dimensional problems.

Once the type of finite element is chosen, the discretization process is started taking into account the following:

1. The intersection between any two finite elements shall be on the edge or a vertex.
2. The size of the element is not important; however, a larger quantity of elements shall be allocated in regions where a bigger field variation is expected.
3. Only one homogeneous medium inside the element is admitted.

Figure 1.5 shows a discretized domain where these requirements can be observed.

1.2.4   STEP 4: SELECTION OF THE INTERPOLATOR FUNCTION

The concept of interpolation is often used when it is intended to estimate the value of a function in a specific point by using the value of the same function in other points around to the one. This selection is closely connected to the type of element used for the discretization of the domain. In this context, we are going to focus on simpler elements, for example, the straight-sided triangle using a 2D case. A simple extension of this concept is also applied for the straight-sided tetrahedron using a 3D case. A simpler case is the one-dimensional linear interpolation where from the function value of a single variable, in two extreme points of a segment (denominated element), its value is estimated on any internal point using a straight line that satisfies its value on the extreme ends of the element, as shown in Figure 1.6.

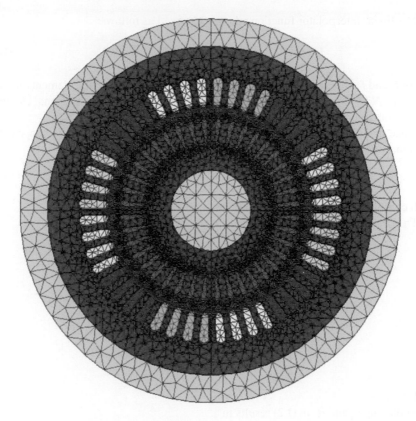

FIGURE 1.5 Mesh for an induction motor.

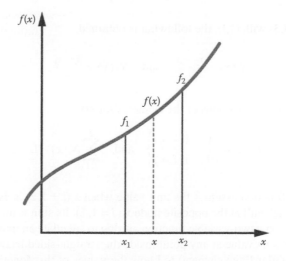

FIGURE 1.6 One-dimensional linear interpolation.

The linear interpolator function can be expressed as follows:

$$f(x) = N_1(x)f_1 + N_2(x)f_2 \quad \text{with } x_1 \le x \le x_2 \tag{1.1}$$

where f_1 and f_2 are the values of $f(x)$ at the end points or nodes of the element and the functions $N_1(x)$ and $N_2(x)$ are named shape functions of the one-dimensional element that are obtained from the simple following procedure:

Since the interpolation is linear, the $f(x)$ function at the interval $x_1 \le x \le x_2$ can be written as following:

$$f(x) = \alpha_1 x + \alpha_2. \tag{1.2}$$

with coefficients α_1 and α_2 to be established.

This function is applied to the nodes of the element, obtaining,

$$f_1 = \alpha_1 x_1 + \alpha_2$$

$$f_2 = \alpha_1 x_2 + \alpha_2$$

so that,

$$\alpha_1 = \frac{f_2 - f_1}{l} \quad \text{and} \quad \alpha_2 = \frac{x_2 f_1 - x_1 f_2}{l}$$

with $l = x_2 - x_1$.

Replacing α_1 and α_2 in (1.2) results in

$$f(x) = \frac{x_2 - x}{l} f_1 + \frac{x - x_1}{l} f_2. \tag{1.3}$$

Identifying (1.3) with (1.1), the following is obtained:

$$N_1(x) = \frac{x_2 - x}{l} \quad \text{and} \quad N_2(x) = \frac{x - x_1}{l}. \tag{1.4}$$

Note that the shape functions of the elements satisfy

$$N_i(x_j) = \begin{cases} 1 & \text{if } i = j \\ 0 & \text{if } i \ne j \end{cases} \quad \text{and} \quad \sum_{i=1}^{2} N_i(x) = 1.$$

So, the $N_i(x)$ function assumes the unit value when $x_i (i = 1,2)$ is calculated at its nodes and becomes null at the opposite node $x_j (j = 1,2)$. Its sum is unitary.

We now focus on a two-variable function aiming to calculate an interpolator function that estimates its value at any point inside the straight-sided triangular element (first-order triangular finite element) to know the values of this function on the element's vertex or nodes.

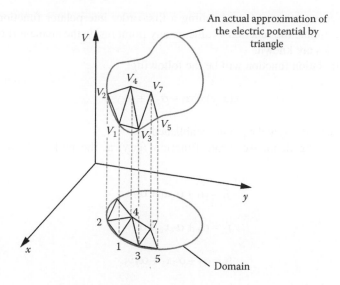

FIGURE 1.7 Two-dimensional linear interpolation.

Figure 1.7 shows a discretized domain for first-order triangular finite elements. Consider that the values of a two-variable function $f_i = f(x_i, y_i)$ are known only at nodes (vertexes) for the discretization. Figure 1.7 also shows, in relief, a surface corresponding to the exact solution of the function $f(x,y)$ and the surface corresponding to the approximated one, formulated with values of the node functions of the finite element mesh and interpolated by planes connecting three values of the function to the three nodes of the element [53,54].

Figure 1.8 presents a two-dimensional linear interpolation applied to a generic triangular element extracted from the finite element mesh.

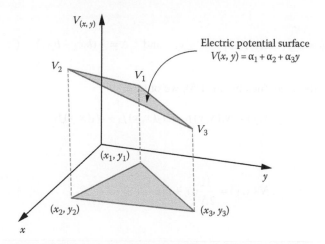

FIGURE 1.8 Two-dimensional linear interpolation for a triangular element.

Our objective consists of establishing a first-order interpolator function, which allow us to estimate the function's value on any point inside the element if their values at the nodes are known.

The interpolation function will be the following:

$$f(x,y) = \alpha_1 + \alpha_2 x + \alpha_3 y \qquad (1.5)$$

with coefficients α_1, α_2 and α_3 to be established.

For calculating them, we wrote function $f(x,y)$ at the nodes of the element to obtain

$$f_1 = \alpha_1 + \alpha_2 x_1 + \alpha_3 y_1$$

$$f_2 = \alpha_1 + \alpha_2 x_2 + \alpha_3 y_2$$

$$f_3 = \alpha_1 + \alpha_2 x_3 + \alpha_3 y_3.$$

The resolution of the three equations, system gives the values of the searched coefficients so that

$$\alpha_1 = \frac{1}{2\Delta}(a_1 f_1 + a_2 f_2 + a_3 f_3)$$

$$\alpha_2 = \frac{1}{2\Delta}(b_1 f_1 + b_2 f_2 + b_3 f_3)$$

$$\alpha_3 = \frac{1}{2\Delta}(c_1 f_1 + c_2 f_2 + c_3 f_3)$$

with

$$a_1 = x_2 y_3 - x_3 y_2; \quad b_1 = y_2 - y_3 \quad \text{and} \quad \Delta = \frac{1}{2}(b_1 c_2 - b_2 c_1). \quad (^*)$$

Replacing these coefficients in (1.5), we obtain

$$f(x,y) = N_1(x,y) f_1 + N_2(x,y) f_2 + N_3(x,y) f_3 \qquad (1.6)$$

where,

$$N_i(x,y) = \frac{1}{2\Delta}(a_i + b_i x + c_i y) \quad i = 1,2,3. \qquad (1.7)$$

---

* The other coefficients a, b, and c are obtained by cyclic rotation of their indexes, and Δ is the element's area.

As for the one-dimensional case, $N_1(x,y)$, $N_2(x,y)$, and $N_3(x,y)$ functions are denominated shape functions of the first-order triangular finite element and satisfy the following conditions:

$$N_i(x_j, y_j) = \begin{cases} 1 & se\ i = j \\ 0 & se\ i \neq j \end{cases} \quad \text{and} \quad \sum_{i=1}^{3} N_i(x,y) = 1.$$

From this result, we can conclude that the ith shape function assumes value 1 at node i and 0 at the other two nodes.

A compact way for representing the interpolator function of the first-order triangular element is

$$f(x,y) = \sum_{i=1}^{3} N_i(x,y) f_i. \tag{1.8}$$

This notation will be used often in the following chapters.

We now discuss three-dimensional problems, which are discretized by volumetric finite elements, such as a tetrahedron.

1.2.5 STEP 5: SELECTION OF THE TYPE OF STUDY

In electrical engineering, engineers try to determine the potential distribution associated to an electromagnetic problem, which can be the electrical potential in cases involving the establishment of an electric field or the magnetic potential vector in cases involving the establishment of a magnetic field, or sometimes both. This situation occurs when both the electric and magnetic fields are present on the device under study, for example, time-dependent electromagnetic phenomena. With these potentials, the magnitudes of the electric and/or magnetic fields searched, through a post-process, are easily obtained, which are the basis for the assessment of the performance of any electric equipment. In high-frequency, time-dependent electromagnetic phenomena, it is common to directly search the distribution of both fields, the electric and magnetic fields as we will discuss next.

In order to obtain a solution for the electromagnetic problem using the FEM, the first measurement consists of characterizing the regions that constitute the device through one (or more) physical properties. Thus, for each finite element, physical properties associated with the phenomenon present on the device should be attributed; for example, if we are interested in studying high-voltage isolators in the electrostatic problem, we need to know the electric permittivity of the materials (ε) and attribute it to the corresponding elements. In case of magnetostatics, the physical property that shall be considered is the magnetic permeability (μ). For electrokinetic problems, electric conductivity (σ) shall be applied. In other problems involving the presence of not only the electric field but also the magnetic field, the three physical properties may be necessary.

1.2.6 STEP 6: INTRODUCTION OF THE BOUNDARY CONDITIONS

The FEM analysis consists, essentially, of the transformation of a system of differential equations describing an electromagnetic phenomenon in an equation system whose solution is close to the solution of a differential equations system at nodes in meshes of finite elements. As with all differential equations, obtaining the solution requires the establishment of the integration constants, which are obtained by imposing the problems' boundary conditions, also called constraints. In the case of the FEM this is not different because when the boundary conditions are not imposed, the equation system resulting from the application of the method, as we will see, does not have a solution. For electromagnetic field problems, the boundary conditions are imposed at the domain limits. There are three kinds of boundary conditions.

1. *Dirichlet conditions*: In this case, the state variable or degree of freedom is known in part or completely at the domain limits. When we work on potentials, it is frequent to have the null potential in some part of the boundary, not only for the imposition of sources connected to the problem, but also for confining the magnetic field inside the domain.
2. *Neumann conditions*: In this case, in a part of the boundary, the electric field (E) is tangential and the magnetic field (B) is normal to the domain limits.
3. *Periodicity conditions*: These types of boundary conditions are applied in structures representing constructive repetition. Electric machines are typical devices to which these types of boundary conditions are applied because the pattern of a magnetic pole of the machine is repetitive.

We will discuss in detail, how these boundary conditions are imposed on the assembly of an equation system, representative of the solution of an electromagnetic field problem by FEM.

1.2.7 STEP 7: SOLUTION OF THE EQUATION SYSTEM

Once the physical properties are attributed to each finite element, we are able to generate equations' system of equal or multiple order of the number of nodes (*number of vertices*) from the domain, that is,

$$[A] \cdot [x] = [b] \tag{1.9}$$

where
   [A] is the global matrix of the equation systems of equal or multiple order of the number of nodes from the mesh of finite elements
   [x] is the vector of unknown or state variable, constituting the scale potentials or components of vector fields
   [b] is the source (or actions) vector, whose elements depend on the intensities of the field sources and the imposed boundary conditions.

Each line of the equation system is the balance conditions required by Maxwell's equations with the appropriate boundary conditions described in the previous step. The details of this procedure form the core of this book, so that we will dedicate a good part of its content to this task. The inconvenience after solving the equations' systems is that we will have a large amount of data for analysis. In general, it is common to generate systems with tens of thousand equations to tens of thousand unknowns, which makes it practically impossible to arrive at any conclusion after the analysis of this data pool. For this reason, we need one more step in this process, which we will discuss later.

An important detail of this equation system is that the largest part of the electromagnetic phenomena generates a symmetric global matrix. Additionally, in all the cases, the referred matrix is sparse, that is, it presents an immense quantity of null elements. As a reference, the proportion between the non-null and null elements of a global matrix of the order of tens of thousands is only around 3%–5%, which justifies the use of some algorithms using only non-null elements of symmetric matrixes.

Note when a data pool of this size is manipulated, advanced techniques of computational programming are required for utilizing the available computational resources to the maximum. Some electromagnetism problems use a nonlinear equations' system where the elements of the global matrix [A] depend on the value of the elements of the unknown vector [x].

In situations like these, which are easily found in electromagnetic phenomena presented on ferromagnetic structures, for example, electric machines, transformers, and other devices, the resolution of problems is made through an iterative process. The number of iterations (number of times where a system of linear equations type $[A][x] = [b]$ needs to be solved) depends on the imposed accuracy of the solution and the level of nonlinearity of the problem [58]. We will discuss in detail this process when we present the basis of magnetostatics using FEM.

1.2.8 STEP 8: EXPLORATION OF RESULTS—POSTPROCESSING

The exploration of results consists of a computational procedure, which uses many of the resources from graphic computation, to have all that data generated from the previous step translated into graphic images for easy understanding (Figure 1.9).

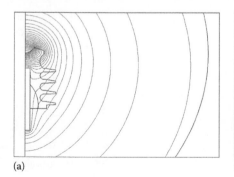
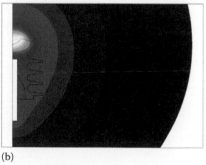

(a)   (b)

FIGURE 1.9 Exploration of results: isolator of a transmission line. (a) Equipotential lines. (b) Color shade electric field.

Some interesting images are

- *Field lines:* The alignment of field lines is important for identifying if the obtained solution is physically correct.
- *Color shades*: This image shows the results using a color gradient image of the domain, from light to more intense, the electric or magnetic field distribution all over the domain, highlighting the regions where their intensity can overcome the operation limits of the material.
- *Graphs*: Graphics displaying the state-variable and/or field variation involved in the study, based on the function of the position, for making easier the understanding of the phenomenon and the corrective decision-making in case of violation of admissible limits depending on the material used on the devices manufacturing.

Besides these images, other values of real interest for the design are obtained after adequate processing. A few worth mentioning are calculations of the electromagnetic effort origin, parameters of the equivalent circuit, and losses of magnetic origin.

Current efforts of research studies are oriented to the optimization and coupling processes with other types of phenomena related to electromagnetic phenomenon, such as magnetic–thermal, magnetic–mechanic couplings, and others.

In this book, we will limit the analysis of electromagnetic phenomena to the initial objectives of this book described earlier to be accessible to both undergraduate students and others who are beginners learning this methodology.

2 Fundamentals of Electromagnetism

2.1 SOURCES OF ELECTROMAGNETIC FIELD

Richard Feynman was one of the greatest scientists of the twentieth century, who won the Nobel prize in physics and the author of one of the most famous works on fundamental physics (*Lectures on Physics*) [16]. He suggested in his book that if civilizations became extinct leaving you as the only human being in contact with a new civilization being born and if you could communicate through one sentence during that time, then that sentence should bear the most important information from humanity. This is the concept behind atom formation, where a set of negative electric charges surround a set of positive charges indefinitely. This information, which is considered to be the first mode of interaction in the field of science, marked the beginning of evolution.

The concept of electric charge is rooted deep in our knowledge, and only after a closer involvement with science was established, we started to understand its origin through atoms without doubting this concept. We were also able to understand the distribution of isolated electric charges in space thereby understanding several physical phenomenon, such as Coulomb's law.

On the other hand, the human body, at a microscopic level, essentially consists of electric charges. Feynman also asserted that if an unbalance of only 10% is found between positive and negative electric charges in our body, the intensity of force between you and the next person, would be sufficient to move the entire Earth!

These electric charges are responsible for the generation of electromagnetic fields. Their state of animation is also relevant; if electric charges are stationary or moving (constituting an electric current), their actions over electromagnetic fields are completely different, as we are going to discuss next. Actions of electric charges in applied electromagnetism are based on the macroscopic conception of the phenomenon.

Based on the macroscopic point of view, the size (and the mass) of the elementary charge is not relevant, that is, we do not differentiate these criteria, for example, an elementary positive charge from a negative one, although the difference in size (and mass) of an electron related to a proton is relevant and exceeds million times. To get an idea about this difference, imagine that a hydrogen atom through a magical process, increases its size until the electronic cloud reaches the size of a cathedral. The electron also appears as a dimensionless point to us and we probably would not see it, as long as the proton has the size of a pearl situated at its center and weighs more than the cathedral.

Based on the microscopic point of view, we are involved with the study of particle physics where other forces are present, everything is in rapid movement, and valid concepts are the ones regarding quantum mechanics. It consists in "seeing" from far the actions generated from microscopic effects and minimization of the difficult level of analysis (I hope you will agree with this!). Thus, as with humanity, it is easier to foresee the crowd's behavior than the individual's.

2.1.1 Volumetric Distribution of Electric Charges

Figure 2.1 shows a volume named with the Greek letter, τ which is surrounded by a closed surface (external surface) named with the Greek letter Σ. We assume that the interior of this volume has electric charges distributed uniformly. This distribution of charges is such that if we select a region around any point of the volume, we can identify the quantity of charge contained in that region, that is, electric charges within it are continuously distributed. Note that, strictly speaking, all charges in this volume do not present a continuous distribution as mentioned earlier, but what we obtain are discretized charges, called punctual charges, as there is a small quantity of universal charge which is, for us, the charge of an electron (-1.6×10^{-19} C).

However, when we analyze the problem under a macroscopic point of view, this detail is not "seen" by us. To have a clear idea of this concept, look for a cloud and try to identify water drops that compound it. It is not possible to have a clear view from the ground level because the view from the ground depicts that water is present at any point of the cloud's volume—this is the macroscopic view that we mentioned. Whereas, if we view the clouds from above and look attentively, we can identify the water drops of which the clouds are composed—this is the microscopic view of the same phenomenon.

Select an elementary volume of τ volume and count the quantity of electric charges inside the volume. Be Δq the net quantity of electric charges contained in the elementary volume $\Delta \tau$.

We define the value ρ_v, denominated volumetric density of electric charges, to get the following relation:

$$\rho_v = \lim_{\Delta \tau \to 0} \frac{\Delta q}{\Delta \tau}. \tag{2.1}$$

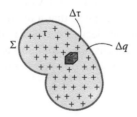

FIGURE 2.1   Volumetric distribution of electric charges.

If the limit exists (it will not exist only if the macroscopic concept is abandoned), it can also be written as

$$\rho_v = \frac{dq}{d\tau} \quad (C/m^3).$$ (2.2)

Inversely, the total amount of electric charges contained in a certain volume through the volumetric integration of the volumetric density of charges in this volume, could be determined as follows:

$$Q = \int_\tau \rho_v \, d\tau.$$ (2.3)

2.1.2 Superficial Distribution of Electric Charges

Figure 2.2 shows a convex surface—named with the letter *S*—which *is surrounded by a closed line (external limit)—named with the letter C*. Let us assume that electric charges are distributed uniformly on this surface. In this situation, it is not possible to apply the concept of volumetric density of charges since it is not possible to isolate an elementary volume of this geometry, because any cross section on this surface produces a line segment. As seen previously, this charge distribution is such that if we select a region around any point on the surface, we can identify the quantity of charges contained in this region, that is, electric charges are continuously distributed, according to the macroscopic concept discussed.

Next select an elementary surface *S* and account the quantity of electric charges present inside. Be Δq the net quantity of electric charges contained on the elementary surface ΔS. We define the value ρ_S, denominated superficial density of electric charges, to get the following relation:

$$\rho_S = \lim_{\Delta S \to 0} \frac{\Delta q}{\Delta S}.$$ (2.4)

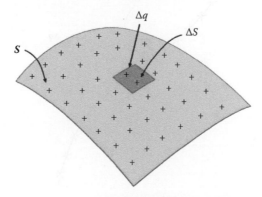

FIGURE 2.2 Superficial distribution of electric charges.

If the limit exists, we get

$$\rho_S = \frac{dq}{dS} \quad (C/m^2). \tag{2.5}$$

Inversely, the total amount of electric charges present on a certain surface could be determined, through the superficial integration of the superficial density of charges over this one, as

$$Q = \int_S \rho_S \, dS. \tag{2.6}$$

2.1.3 LINEAR DISTRIBUTION OF ELECTRIC CHARGES

Figure 2.3 shows a line L where we have electric charges distributed throughout. In this situation, it is not possible to apply concepts of both volumetric and superficial densities of electric charges, because we cannot isolate a volume or an elementary surface of this geometry since any cross section on this surface forms a point.

Similarly, as mentioned earlier, this distribution of charges is such that if we select a region around any point of the line, we can identify the quantity of charges contained in that region, that is, electric charges in the line are continuously distributed, according to the macroscopic concept discussed.

Select an elementary segment from line L and evaluate, as done previously, the quantity of electric charges inside it. Be Δq the net quantity of electric charges contained in the elementary segment Δl. We define the value ρ_l, denominated linear density of electric charges, to get the following relation:

$$\rho_l = \lim_{\Delta l \to 0} \frac{\Delta q}{\Delta l}. \tag{2.7}$$

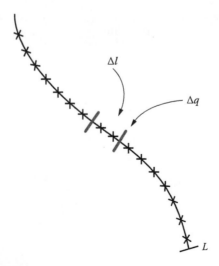

FIGURE 2.3   Linear distribution of electric charges.

If this limit exists, it results in

$$\rho_l = \frac{dq}{dl} \quad (C/m). \tag{2.8}$$

Inversely, the total amount of electric charges contained in a line could be determined, through the integration of the linear density of charges over this one, as following:

$$Q = \int_L \rho_l \, dl. \tag{2.9}$$

2.1.4 DISCRETE ELECTRIC CHARGES

In many cases, it is convenient to consider electric charges concentrated at a certain point, mainly when the aim is to evaluate values depending on the charge in distant regions. These are called discrete electric charges where it is not possible to define any density of charges.

2.1.5 CURRENT DENSITY VECTOR

Electric charges can also have movement, which is termed as electric current. The electric current, which is a scalar quantity, cannot be used to analyze, as we will discuss in later chapters, the electromagnetic phenomena to which the direction (and sense) of the current shall be considered. For this reason, we need to establish a vector field associated with the electric current.

Figure 2.4 shows a conductor conducting electric current (i). Current lines represent the characterization of the electric current's flux inside the conductor. Lines are drawn so that the quantity of current contained between two consecutives of them is maintained constant. If the same figure were three dimensional, we would have current tubes conducting a constant current in its interior.

Isolate an elementary current tube as shown in Figure 2.5. Δi is considered as the intensity of current flow through the elementary tube and ΔS_n as the cross section of the tube, that is, the normal section to the current lines.

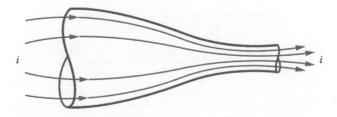

FIGURE 2.4 Current flux on conductors.

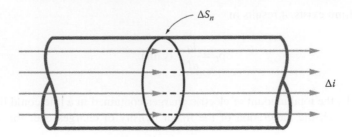

FIGURE 2.5   Elementary current tube.

Defining the *current density vector* $\vec{J}$, with following characteristics:

$$\text{Module: } J = \lim_{\Delta Sn \to 0} \frac{\Delta i}{\Delta Sn} \quad (\text{A/m}^2), \tag{2.10}$$

Direction: Tangent to the current lines,

Sense: Positive if agreeing with the current sense.

Considering the elementary current tube, let us select any elementary surface ΔS as shown in Figure 2.6.

Of course, the current intensity crossing the ΔS and ΔS_n surfaces is the same

$$\Delta i = J \Delta S_n = J \Delta S \cos \alpha, \tag{2.11}$$

where α represents the angle established between the normal straight ΔS and the current lines, as shown in Figure 2.7.

The result obtained in Equation 2.11 reminds us of a dot product of two vectors. As the $\vec{J}$ vector is already well characterized, we can define vector $\Delta \vec{S}$ so that its module is equal to the elementary area ΔS, normal to the surface, and with an arbitrary sense.

Thus, we can write

$$\Delta i = \vec{J} \cdot \Delta \vec{S}. \tag{2.12}$$

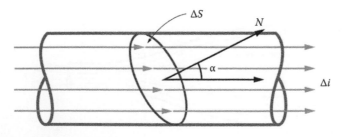

FIGURE 2.6   Elementary current tube.

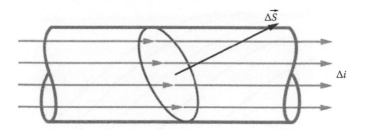

FIGURE 2.7 Elementary area vector.

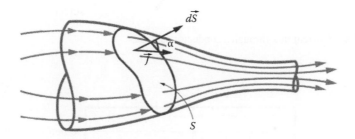

FIGURE 2.8 Electric current inside a conductor body.

Supposing now that is wished to calculate the intensity of current crossing any surface extracted from a conductor body flown by the electric current, when the current density vector $\vec{J}$ is known inside it, Figure 2.8.

An alternative consists in subdividing the surface S, as shown in Figure 2.8, into a very large number of tiny surfaces, applying Equation 2.2 to each one of them, and then adding up the values obtained to calculate the total current crossing the surface. If a large number of small surfaces are obtained from the subdivision of S, the total current can be expressed as

$$i = \int_S \vec{J} \cdot d\vec{S}. \tag{2.13}$$

2.1.6 CURRENT SUPERFICIAL DENSITY VECTOR

Electric charges also show movement over a surface where the cross section is a line. Although physically it is not possible, because a conductor surface with this characteristic does not exist, but this is possible through engineering if the cross section of the conductor is extremely smaller than dimensions involved in the problem.

The characterization of current density vector, with same requirements as mentioned earlier, is not possible since we cannot make a non-null cross section on this surface through which the current flows.

FIGURE 2.9   Current flux through a conductor surface.

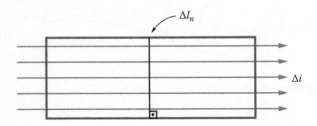

FIGURE 2.10   Current filament.

Figure 2.9 shows a conductor surface generating an electric current i. Current lines represent the characterization of the current flow.

Isolate an elementary current band, as shown in Figure 2.10, and consider Δi as the intensity of current flowing through this band.

If Δl_n is the cross section of this band, that is, the normal segment to the current lines, the *current superficial density vector* designated by $\vec{J}_l$ shows following characteristics:

$$\text{Module: } \vec{J}_l = \lim_{\Delta \ln \to 0} \frac{\Delta i}{\Delta l_n} \quad \text{(A/m)}, \tag{2.14}$$

Direction: Tangent to the current lines,

Sense: Positive if agreeing with the current sense.

Similarly, the total amount of current crossing any line supported on the surface, as shown by Figure 2.11, is given by

$$i = \int_L \vec{J}_l \cdot \vec{dl}. \tag{2.15}$$

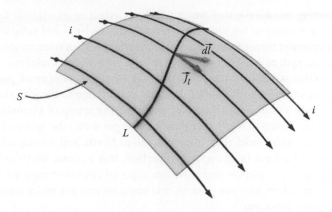

FIGURE 2.11 Conductor surface flown by electric current.

Note that the elementary segment vector $\vec{dl}$ is always normal to the line at any point.

Observation: When these densities are constant, the associated distribution is considered to be uniform.

2.2 FIELD VECTORS

Static or moving electric charges alter the space properties that involve them, and these properties are characterized by actions produced in other static or moving charges. Thus, they have been given the generic name "fields," because they act in a well-defined region from space [15].

2.2.1 THE ELECTRIC FIELD VECTOR

Electric field is an additional property acquired by a space when an electric charge is placed in its surroundings. The question is which property is this one?

According to a majority of physicians, this property is the action of an acting force over another charge placed in this space that depends directly on its quantity and also on the intensity of the field present, that is,

$$\vec{F} = q\vec{E}. \tag{2.16}$$

Dimensionally, the electric field is measured by

$$[E] = \frac{[F]}{[q]} = \frac{N}{C}.$$

Another dimension based on the International System of Units equivalent to the prior one, which is often used for measuring the electric field, is the V/m (Volt/meter), which reason we will discuss later in this chapter.

It is interesting to observe that we can feel the effect of an electric field in our own body in the vicinity of big high-tension transmission lines and substations; such sensation is perceived through our skin and hair, which can, in some cases, make our hair stand at one end, as it happens in science fairs with young students.

This sensibility is because our body consists of equal number of positive and negative electric charges, which suffer different actions due to the electric field surrounding our body. For example, when we are under the action of a constant electric field directed from South to North, our positive charges suffer the action of a force in a similar direction to the field, that is, from South to North, and our negative charges suffer the action of a force in an opposite direction, that is, from North to South.

Fortunately, electric fields to which we are exposed are not enough for breaching the existing connections between positive and negative charges and it results in only a short discomfort sensation.

This effect is explored for performing an important domestic task; otherwise, if the electric field that was directed from South to North changed the direction, our electric charges would also feel this change and they would be inversely orientated. If this procedure was continuous, our electric charges would be continuously moving, searching for field orientation, resulting in a heating due to abrasion between the moving molecules. This phenomenon is observed in microwave ovens while cooking food, because this equipment's electric field oscillates with an elevated frequency producing enough abrasion between the molecules for an efficient heating of food.

2.2.2 THE MAGNETIC FIELD VECTOR

...iron can be attracted by a stone which Greeks originally call Magneto, because it's original from Magnesia lands, inhabitants from Magnesia in Thessaly...

This text—written by Lucretius in 100 BC—was the first evidence for the existence of a physical phenomenon, which was responsible, almost after two millennia, for the tremendous development experienced by humanity. That statement regarding Magnesia asserts that some rocks of that region had the property of attracting other similar stones. Observation of the phenomenon gave birth to a new scientific study, named magnetism, which constituted in analyzing the origin and effect of the force field. Based on initial observations, this force field was later named magnetic flux density (or simply flux density vector). We will discuss the reason behind the name later.

On the other hand, in China, the compass was discovered for exploring the earth's magnetic field for guiding sailors. Just after some centuries, the force field produced by Greek stones (later named lodestone) was identified, with similar properties to that of the earth's magnetic field.

Pierre de Maricourt—christened as Peter Peregrinous by the Pope who was crusade army's engineer from Count of Anjou—was credited for creating the science of magnetism, and he had experiences working with lodestone as mentioned in a letter addressed to a friend on August 8, 1269. Pierre de Maricourt was the one who coined the word pole for characterizing different behaviors of each side of a magnet. The letter was written in Lucera/Sicily during a crusade. Considered as the first experimental scientist of the world, he was the first to study magnets reasonably.

The first book on magnetism was written 330 years later, in 1600, by William Gilbert, named *De Magnete*, in Latin.

Magnetism during that period was treated as an independent science, having no connection with electricity studies, which had conflicts with studies of Alessandro Volta, Charles Coulomb, and other renowned researchers. However, the best was about to happen.

Hans Christian Oersted was the older son of a poor old pharmacist born in Rudköbing, at the Baltic island of Langland, Denmark, on August 14, 1777. Excited with the discovery of the voltaic cell of Volta, he used one of them for producing the electric current circulation in alkaline acids. While at the university, Oersted helped a professor to organize the institution's pharmacy, which he used for his experience. This work fetched him a scholarship, which constituted of a trip to the main European laboratories of that time. In 1806, he became a physics professor at Copenhagen University. Although Coulomb and Ampère (who he came to know later) believed that magnetism and electricity were different sciences, which could not be related, Oersted, however, believed that magnetism existed in all bodies like electricity did. Later he wrote that he strongly believed that the same force from electricity could produce the effects of magnetism, For example, boats hit by lightnings experiencing their compasses being demagnetized.

He always believed that the electric current passing through a fine wire conductor could generate magnetic effect after becoming incandescent and emit light; to prove this he made an experiment where the electric current would pass through a fine platinum wire to see its effect on the compass needle. It was a simple test, but he lacked enough time for setting it up before its presentation. Afraid of any mishap, he decided to cancel this part of the presentation.

During his presentation, however, he mentioned again the relation between electricity and magnetism, but he could not resist from performing the experiment for the audience. The compass was placed close to the wire and on passing current through the wire the needle slightly deflected.

A painting of Louis Figuier portrays this moment showing Oersted and a student of his, who acted as an assistant. This student was who first observed the phenomenon, because the wire had not been sufficiently heated for emitting light and producing magnetism—as previously foreseen by Oersted.

The scientific historian George Staton stated that this experiment was considered "between the most memorable in all science's history." On July 1820, he sent a four-page article about this discovery to several scientific journals and in a few weeks the world acknowledged that electric current could generate magnetism. This marked the beginning of the most fascinating science of all times, the electromagnetism. Soon after Oersted published his observation, its repercussion had a huge effect on the scientific community at that time led, among others, by Ampère. Inspired by Oersted's experiment, Ampère created electrodynamics (electricity in movement), between 1820 and 1827. In 1821, the work of Biot and Savart was noticed, as they quantified the magnetic flux density field produced by an electric current distribution.

Thus, almost after two millennia, magnetic flux density field was identified in the field of modern science. For quantifying it, science made what is always made when a forces field is identified, that is, it was attributed an intensity to the field based on already known quantities such as mass, length, time and electric charge. In case of magnetic field, it was observed that this one exerts a force over a conductor run by the electric current represented by [14,20]

$$\vec{F} = \int_L i d\vec{l} \times \vec{B}. \tag{2.17}$$

A dimensional analysis of Equation 2.17 indicates that the magnetic flux density vector field dimension is represented as

$$[B] = \frac{N}{A \cdot m},$$

where the International System of Units attributed the name Tesla for identifying it, therefore

$$\text{Tesla (T)} = \frac{N}{A \cdot m}.$$

The CGS system attributed the name Gauss (G) as magnetic flux density field unit, and it is easy to show that

$$1 \text{ T} = 10^4 \text{ G}.$$

However, a question started to perturb the scientific community after Oersted's discovery. If electricity can produce magnetism, could magnetism produce electricity? Would nature be so asymmetric at this level? We will soon answer this question.

2.2.3 DISPLACEMENT VECTOR

During eighteenth century, experiences with electric field considerably advanced and, in one of the most important experiences, it was announced that the electric field flux over a closed surface is directly proportional to the total amount of electric charges contained within this surface.

Let us understand—accurately—the meaning of the electric field vector flux over a closed surface.

Back then certain quantities of field lines were associated to the intensity of the electric field and, through a judicious process, lines crossing the surface were counted, so it was observed that if the quantity of electric charges within the surface varies, the quantity of lines crossing it also varies proportionally. This law—named Gauss's law of electrostatics—is mathematically expressed as

$$\oint_\Sigma \vec{E} \cdot d\vec{S} = \frac{Q}{\varepsilon_0}. \tag{2.18}$$

The constant of proportionality, written as $1/\varepsilon_0$, was identified as a specific property of the medium involved—which in this case was the air—so for this reason the constant ε_0 was named *electric permittivity of the air* (or of the vacuum).

There are no doubts that the experiment is not possible having a dielectric medium different from air due to difficulty in measurements. However, the result is generalized, attributing a dielectric property for a medium different from the air's, forming a concept of electric permittivity of the medium characterized by the Greek letter ε. Experiences also showed that the electric permittivity of air is the lower electric permittivity possible, reason why the relative permittivity ε_r was defined and established as $\varepsilon_r = \varepsilon/\varepsilon_0$. Thus, Gauss's law was represented as following:

$$\oint_{\Sigma} \vec{E} \cdot d\vec{S} = \frac{Q}{\varepsilon}. \tag{2.19}$$

The evolving technology required again a change in Gauss's law for situations where the electric permittivity of the medium is depended on the existent electric field. Indeed, the only possible change was the insertion of the electric permittivity inside the integral of Gauss's law, for considering the possibility of the electric permittivity being dependent on the electric field.

This is very common, because it is directly connected to the alignment of atoms of the material with the electric field imposed, which we discuss next. Thus, the almost final form of Gauss's law became

$$\oint_{\Sigma} \varepsilon\vec{E} \cdot d\vec{S} = Q. \tag{2.20}$$

The dimensional analysis of Equation 2.20 concludes that the dimension of the electric permittivity of a medium is given by

$$[\varepsilon] = \frac{C}{\dfrac{V}{m}m^2} = \frac{F}{m},$$

where

$$F\,(\text{Faraday}) = \frac{C\,(\text{Coulomb})}{V\,(\text{Volt})}.$$

A small change in this law still occurs, simply replacing $\varepsilon\vec{E}$ by $\vec{D}$ in Equation 2.20 resulting in

$$\oint_{\Sigma} \vec{D} \cdot d\vec{S} = Q. \tag{2.21}$$

Vector $\vec{D} = \varepsilon\vec{E}$ is called the displacement vector, with a dimension of C/m². The reason for the name is discussed later. However, it is noted in Equation 2.21 that the flux of the displacement vector over a closed surface is equal to the total amount of electric charges contained within the surface. Therefore, in electromagnetic problems it is advised to replace the electric charge by an electrical displacement flux, because it is more convenient to manipulate the displacement vector than the electric charge. It is also noted that the medium does not affect this integral (note that in Equation 2.21 no quantity appears based on the permittivity of the medium). However, it is important to highlight that the flux (integral of closed surface) of the displacement vector over a surface does not only depend on the medium and the displacement vector but also on the medium where the electric field is present.

2.2.4 MAGNETIC FIELD INTENSITY VECTOR

During nineteenth century, the number of researchers involved in Ampere's work increased , who continued obtaining new laws regarding the science of electromagnetism. The main law, which was obtained in the first half of the century, named as Ampère's circuital law was a homage to the genius.

Studies of these researches showed that *circulation—closed path integral—of the magnetic field was directly proportional to the total amount of linkage current by the closed path over which the circulation was performed.* Questions are: Why these researches concentrated on obtaining a circulation and not on a surface integral such as Gauss's law? What does it mean by *linkage* current by a closed path?

Let us discuss these questions starting with the second one. Figure 2.12 helps us to understand this concept.

Current linkage for a closed path is the scalar quantity equal to the electric current through any surface bounded by the path.

Thus, if we apply this concept, linkage currents by the closed path shown in Figure 2.12 are i_1, i_2, i_4 (three times). Note that i_3 is not a linkage current because it does not cross *any* surface laid on C.

A sign is also attributed to the linkage current and for that the closed path C shall be orientated—arbitrarily—as shown in Figure 2.13.

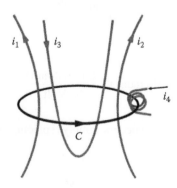

FIGURE 2.12   Linkage current by a closed path.

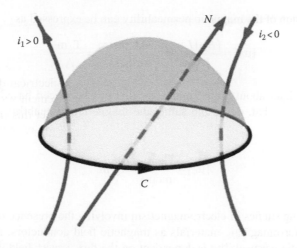

FIGURE 2.13 Linkage current: convention.

Establishing the orientation of the closed path, the normal one is orientated, applying the right-hand rule, at any point on the surface so that the hand's fingers, except the thumb, follow the orientation of the closed path resulting in the direction of the agreeing normal with the thumb direction. Figure 2.13 shows the normal orientation in a generic point of that surface so that we can identify if it is in compliance with the previous hypothesis.

With these orientations established, we account as positive the linkage current with the closed path which direction is agreeing with the normal direction to the surface and negative in opposite case.

Thus, for the current distribution shown in Figure 2.12, the total amount of linkage currents (i_t) by that closed path is given by

$$i_t = i_1 - i_2 - 3i_4.$$

Considering the first question, efforts are concentrated in obtaining the field circulation over a closed path for identifying if the magnetic flux density field vector was or was not conservative—fields for which circulation is null—as the case of a static electric field. We will explain this while discussing the concept of voltage later.

Finally, the first mathematical representation of Ampère's circuital law is given by

$$\oint_C \vec{B} \cdot d\vec{l} = \mu_0 i_t. \qquad (2.22)$$

The constant of proportionality μ_0 was identified as a physical property of the involving medium and, as the initial experiences were performed having the air as the involving medium, μ_0 was considered as the magnetic permeability of air (or vacuum).

The dimension of the magnetic permeability can be expressed as

$$[\mu_0] = \frac{[B] \cdot [L]}{[A]} = \frac{\text{Tesla} \cdot \text{metro}}{\text{Ampère}} = \frac{\text{T} \cdot \text{m}}{\text{A}}.$$

We discuss later about the permeability that can be placed based on the magnetic flux unit, so that, doing the same, the magnetic permeability dimension is expressed as

$$[\mu_0] = \frac{\text{Henry}}{\text{metro}} = \text{H/m}.$$

With evolving studies of electromagnetism involving the presence of matter, that is, the use of ferromagnetic materials as magnetic field conductors, it is observed that the magnetic permeability is dependent on the flux density field, for which the expression (2.22) is rewritten in following way:

$$\oint_C \frac{\vec{B}}{\mu} \cdot d\vec{l} = i_t. \tag{2.23}$$

The current version of Ampère's circuital law is obtained replacing $\vec{B}/\mu$ by $\vec{H}$ in Equation 2.23 resulting in

$$\oint_C \vec{H} \cdot d\vec{l} = i_t. \tag{2.24}$$

The vector $\vec{H} = \vec{B}/\mu$ with a dimension of A/m called as magnetic field intensity vector (or simply magnetic field vector) is an analogy to the electric field vector, which is also known as electric intensity vector.

In this case, it is also observed that the circulation of the magnetic field vector is equal to the total amount of linkage currents present in these closed paths, and it can replace them very well on calculating. Note that—similar to Gauss's law—the Ampère's circuital law, which is expressed in terms of the magnetic field vector $\vec{H}$, does not depend on the medium but depends only on the total amount of linkage currents by the closed path. It is convenient, however, to highlight that an $\vec{H}$ circulation on the closed path does not only depend on the medium and the magnetic field vector $\vec{H}$ but also on the involving medium of currents.

2.2.5 THE ELECTRIC POLARIZATION VECTOR

Figure 2.14 shows a cross section of a parallel metal plate of a capacitor in which the interior is partially inserted into an isolating material with an electric permittivity ε When a voltage is applied between the capacitor's plates, an electric field

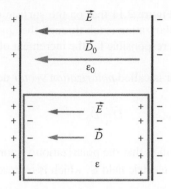

FIGURE 2.14   Parallel plate capacitor—polarization vector.

is established between them. It is demonstrated that the value of the electric field
is equal to $E = V/d$, where V is the voltage applied between the plates and d is the
distance between them, if the region is air or a dielectric medium.

Thus, on the region with air, the displacement vector is such that

$$\vec{D}_0 = \varepsilon_0 \vec{E}.$$

Whereas on the dielectric region, the displacement vector is

$$\vec{D} = \varepsilon \vec{E}.$$

In conclusion, the value of the displacement vector increases due to the presence
of a dielectric, because $\varepsilon > \varepsilon_0$. This increment is due to the deformation observed by
an atom of a dielectric in the presence of a field, which increases the electric field, as
shown in Figure 2.15.

Figure 2.15 shows that the deformation of an atom due to a displacement of the
electronic cloud with regard to the nucleus corresponds to a charge separation. This
is because the center of positive charges (protons) does not coincide anymore with
the center of negative charges (electrons), generating electric dipoles—two opposite
charges very close to each other. This phenomenon is called dielectric polarization,

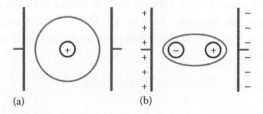

(a)   (b)

FIGURE 2.15   Atom deformation due to the electric field (a) without the action of the electric field (b) under the action of electric field.

and it has been shown in Figure 2.14 that on the surface of the dielectric, next to the capacitor's plate, the polarization is felt through the appearance of an additional superficial charge, which is responsible for the increment of the displacement vector from $\vec{D}_0$ to $\vec{D}$.

This displacement vector is called *polarization vector* designated by $\vec{P}$, so that

$$\vec{D} = \vec{D}_0 + \vec{P}. \tag{2.25}$$

Experimentally, it is verified that the polarization vector $\vec{P}$ on linear materials is directly proportional to the electric field $\vec{E}$, which is represented as

$$\vec{P} = \varepsilon_0 \chi_e \vec{E} \quad (\text{C/m}^2)$$

where χ_e is the *electric susceptibility* of the medium. Replacing this value in Equation 2.25 results in

$$\vec{D} = \varepsilon_0 (1 + \chi_e) \vec{E}. \tag{2.26}$$

Therefore, the electric permittivity of the medium can also be rewritten as following:

$$\varepsilon = \varepsilon_0 (1 + \chi_e).$$

2.2.6 THE MAGNETIZATION VECTOR

To understand magnetization vector $\vec{M}$ we consider a coil with N_{turns}, uniformly distributed along a toroid, as shown in Figure 2.16.

Applying Ampère's circuital law to a closed path within the toroid, as indicated in Figure 2.16, we get

$$\oint_C \vec{H} \cdot d\vec{l} = Ni,$$

where Ni is the total amount of the linkage current within the closed path C.

As the lines from the magnetic field vector $\vec{H}$ always form closed paths, also added to the fact that $\vec{H}$ is constant over this path, so the previous expression can be rewritten as following:

$$H \times \text{closed path perimeter} = Ni$$

or as

$$H 2\pi r = Ni.$$

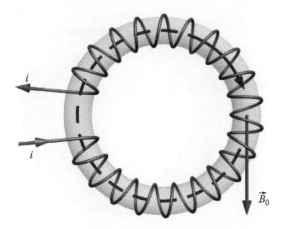

FIGURE 2.16   Toroid.

Note that $\vec{H}$ and $\vec{dl}$ are aligned, because both are tangent to the closed path C. It therefore results in

$$H = \frac{Ni}{2\pi r}.$$

If the material of the toroid is not magnetic, that is, its magnetic permeability is equal to that of air, then the magnetic flux density field inside it is

$$B_0 = \mu_0 \frac{Ni}{2\pi r},$$

where μ_0 is the air's magnetic permeability. If, however, the material of the toroid is a magnetic material with magnetic permeability bigger than that of the air, the magnetic field, in its interior, will get an increment ΔB, that is,

$$\vec{B} = \vec{B}_0 + \Delta\vec{B}$$

or as

$$\vec{B} = \mu_0\vec{H} + \Delta\vec{B}.$$

Dividing both by μ_0 we get

$$\frac{1}{\mu_0}\vec{B} = \vec{H} + \frac{1}{\mu_0}\Delta\vec{B}.$$

The vector

$$\vec{M} = \frac{1}{\mu_0}\Delta\vec{B} \quad (\text{A/m})$$

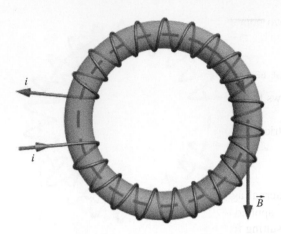

FIGURE 2.17   Toroid: magnetic material $\mu > \mu_0$.

is called *magnetization vector*, which is represented as (Figure 2.17)

$$\frac{1}{\mu_0}\vec{B} = \vec{H} + \vec{M}.$$

Therefore,

$$\vec{B} = \mu_0(\vec{H} + \vec{M}). \tag{2.27}$$

In linear medium, the magnetization vector $\vec{M}$ is directly proportional to the magnetic intensity vector; thus,

$$\vec{M} = \chi_m \vec{H}$$

where χ_m is magnetic susceptibility of the medium.

Replacing its value in Equation 2.27, we get

$$\vec{B} = \mu_0(1 + \chi_m)\vec{H}$$

Since $\vec{B} = \mu\vec{H}$, we get

$$\mu = \mu_0(1 + \chi_m). \tag{2.28}$$

2.3   QUANTITIES ASSOCIATED WITH FIELD VECTORS

The resolution of electromagnetic problems needs some quantities associated with electromagnetic fields. Some of these quantities are frequently used than the vector fields, because it is possible to measure them by instruments. For example, voltage drop between two points in a circuit is measured by a voltmeter in some situations. Frequently, we do not need to know the electric or the magnetic field that creates them.

2.3.1 Voltage between Two Points

Figure 2.18 shows two points A and B located in a region surrounded by an electric field $\vec{E}$.

With the electric field, we are able to evaluate the work involved in dragging the charge q on the path AB, which displacement direction is opposed to the direction of the electric field. Thus, it is necessary the presence of an external agent for taking the charge between that points, because the action of the electric field is opposite to what we propose.

Thus, the external agent that drags the charge should apply a force $\vec{F}_{ext}$ greater than the force $\vec{F}$ applied by the electric field on the charge.

The force resulting from the composition of $\vec{F}_{ext}$ and $\vec{F}$ should have a component tangent to the line of the charge movement.

In addition, if no dragging is required to move the charge along the path AB, that is, the movement is very slow, it implies that the time necessary for dragging the charge from A to B is virtually infinite.

In these cases, we can assert that the force resulting in the composition of $\vec{F}_{ext}$ and $\vec{F}$ is practically null, for which $\vec{F}_{ext} = -\vec{F}$.

Considering our problem, the work done in dragging the load from A to B is given by

$$\tau = \int_{A-B} \vec{F}_{ext} \cdot d\vec{l}. \tag{2.29}$$

Considering an almost stationary condition, the previous equation can be written as following:

$$\tau = -\int_{A-B} \vec{F} \cdot d\vec{l} = -\int_{A-B} q\vec{E} \cdot d\vec{l}. \tag{2.30}$$

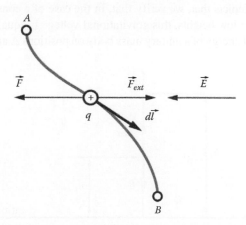

FIGURE 2.18   Points in an electric field region.

We can already define the voltage between both points A and B that is defined as the work performed for dragging a unit charge (work by unit charge) from point A to B, and it is represented as following:

$$V_{AB} = \frac{\tau}{q} = -\int_{A-B} \vec{E} \cdot d\vec{l}. \tag{2.31}$$

A dimensional analysis of Equation 2.31 shows us that the unit of the voltage is *Joule/Coulomb* (*J/C*), which was named *Volt* or simply *V* in homage to Alessandro Volta, the inventor of voltaic battery.

Observe that, at this point we can justify why the electric field unit can also be expressed as V/m, as discussed earlier.

A mechanical analogy may be helpful in understanding the voltage. At the gravitational field, we can define a gravitational voltage between two points, related to heights, as shown in Figure 2.19 [17].

For such, we calculate the work performed for taking a mass m from a point A (lower) to a point B (higher), in a stationary way, which is represented as

$$\tau = -\int_L \vec{P} \cdot d\vec{l} = -\int_L m\vec{g} \cdot d\vec{l}, \tag{2.32}$$

where $\vec{P}$ is the weight of the mass body m. Thus, by analogy, we define the gravitational voltage as the work performed for elevating a unit mass body from point A to B, or as

$$V_{AB} = \frac{\tau}{m} = -\int_L \vec{g} \cdot d\vec{l}. \tag{2.33}$$

Observe that the role of the electric field here is performed by the gravitational field $\vec{g}$. Further analysis depicts that, we verify that, in the case of a constant gravitational field, which occurs in low heights, this gravitational voltage is equal to the difference between the potential energy of a unitary mass body on positions A and B, respectively.

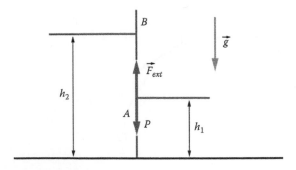

FIGURE 2.19 Gravitational voltage.

Although the voltage was defined from a charge dragged between two points on an immerse region of an electric field, this quantity is independent of the presence of charge, it is a function of both position of the points and intensity of the electric field, and it gives an indication of the capacity of performing the work in this region. In the same way, the gravitational voltage between two points also exists, independent of the presence of a body in the region.

A surface is equipotential when the voltage between any two points is null. It is easy to verify that the integration between two points belonging to that surface is normal to the electric field (note that $\vec{E}$ and $\vec{dl}$ are orthogonal), as shown in Figure 2.20.

Figure 2.20 shows two very close equipotential surfaces. For points A and C of these surfaces, the expression (2.31) can be written as following:

$$\Delta V = -\vec{E} \cdot \vec{\Delta l}$$

or as

$$\Delta V = -E \cdot \Delta l \cos \alpha.$$

Note that the direction that results in smaller Δl is the one normal to surfaces, where $\alpha = 0$ as shown in the figure indicated by the DA direction. Note also that the elementary segment on the DA direction (Figure 2.20) is exactly equal to the projection of any elementary segments that ends are in a point A and in any other point of the other equipotential surface.

This direction is denominated field gradient since through it the potential variation by length unit (i.e., electric field) is maximum. Thus, given an equipotential surface, the electric field is always normal to it.

By analogy with mechanic, if equipotential surfaces are closed path lines of a land, the gradient will be exactly the rain water way flowing over the land.

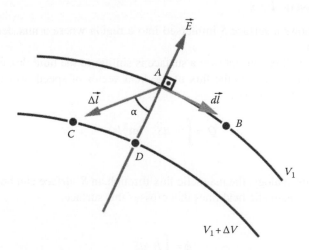

FIGURE 2.20 Equipotential surfaces.

For this reason, the previous expression is represented as

$$E = -\frac{\Delta V}{\Delta l \cos\alpha}$$

or as

$$\vec{E} = -\nabla V. \tag{2.34}$$

From analytical geometry, it has been demonstrated that

$$\nabla V = \frac{\partial V}{\partial x}\vec{u}_x + \frac{\partial V}{\partial y}\vec{u}_y + \frac{\partial V}{\partial z}\vec{u}_z. \tag{2.35}$$

Observe that V function is electric potential function and its definition from partial derivatives require, for its uniqueness, the attribution of a referential. In this case, it is enough to attribute an arbitrary value of potential to any point of the region where the field is present, because if $V' = V + cte$, it results in

$$\nabla V' = \nabla V.$$

The electric potential function is such that the difference of its values calculated at two different points is numerically equal to the voltage between these two points, which can, in some cases, be measured through instruments. Another way of describing it is that the potential function shows the behavior of the voltage between any point and the chosen reference.

2.3.2 MAGNETIC FLUX

Figure 2.21 shows a surface S immersed into a region where a flux density vector field is present.

The magnetic flux concept over a surface is similar to the fluid flux concept over a section, which represents the flux of the field vector of speed over this section, mathematically expressed by

$$Q = \int_S \vec{v} \cdot d\vec{S} \quad (\text{m}^3/\text{s}). \tag{2.36}$$

Through this analogy, the magnetic flux through an S surface can be understood as a quantity of magnetic field lines that crosses this surface.

Thus

$$\phi = \int_S \vec{B} \cdot d\vec{S}. \tag{2.37}$$

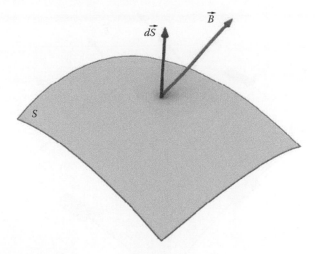

FIGURE 2.21 Magnetic flux.

A dimensional analysis of Equation 2.37 shows that the dimension of the magnetic flux is

$$[\phi] = \mathrm{Tm}^2.$$

Using the fundamental units, we get

$$[\phi] = \frac{\mathrm{Nms}}{\mathrm{C}}.$$

According to the International System of Units, the name *Weber* (Wb) was attributed to the magnetic flux unit, and so magnetic flux density field, as a direct consequence of Equation 2.37, which can also be expressed as the unit (Wb/m²).

The magnetic flux density field can also be understood as magnetic flux by unit of area, which is the reason why we find—frequently—the magnetic induction field denominated by *magnetic flux density* as described in this book.

2.3.3 CONCATENATED MAGNETIC FLUX

Supposing that the border of the S surface from Figure 2.22 is the limit of an electric circuit built with a single grid and immersed into a region where a distribution of flux density field is present. In this case, the magnetic flux that crosses this circuit, also denominated linkage flux, is equal to the magnetic flux calculated by Equation 2.37.

It occurs, however, that we can have circuits constituted by different over placed plans, as the case of coils constituted by several turns ($N_{\mathrm{turns}} = 1$ circuit), subject to a single magnetic flux that crosses them, as shown in Figure 2.23.

Indeed, for evaluating the linkage magnetic flux, we consider the number of times that it crosses the circuit for getting its total. Thus, in a case where we

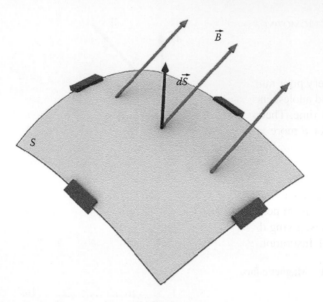

FIGURE 2.22   Linkage magnetic flux with the circuit.

FIGURE 2.23   Concatenated flux with N_{turns} of the coil.

will have a coil with N_{turns}, very close to each other, the linkage flux with this coil is represented as

$$\lambda = N\phi. \tag{2.38}$$

The dimension of the linkage flux is the same as the magnetic flux; however, for differentiating them, a denomination *Weber-turn* (*Wb-turn*) was adopted.

2.3.4   Electromotive Force

2.3.4.1   History

An excellent bookbinder who was popular in the city of London during that time, born into a very poor family. Reports suggest that he suffered from starvation in his childhood and adolescence. He worked 16 h a day, although this workload was very common that time. Therefore, he had no time to go to school, which was reserved, essentially, for a more privileged class. In addition, life contingences led him, as bookbinder, to get in touch with a different culture; he read all the works he had to bind. One day, a client offered him tickets for watching a series of conferences of Sir Humphry Davy (1778–1829), president of The Royal Institution of London—maybe the first public research laboratory of the history—who discovered several chemical elements and gas properties. He wrote down everything that was said and duly bidden those notes, giving them as a gift to Sir Humphry Davy.

The Royal Institution was a research center that counted on the collaboration of several famous scientists. He had a close relationship with The Royal Society of London, the most ancient scientific society on earth, had as president, among others, Isaac Newton, Edmund Haley, and Richard Feynman. When one of his collaborators was fired, Davy remembered the boy who gifted him those notes and so he hired him as his assistant. The boy started to live in the institution's laboratories, immersed in studies. He studied gas properties and he was interested in the study of electricity. After a few years, Michael Faraday—that is his name—who did not know mathematics because he had no regular courses discovered the law of magnetic induction on August 29, 1831 and changed the science invention of electrical engineering.

The phenomenon is very simple and for understanding it correctly, including conventions of signs involved, let us assume a conductor ring, as shown in Figure 2.24, concatenated by a time-dependent magnetic field.

The positive direction of the current is obtained applying the right-hand rule to the closed path, placing the thumb to the agreeing direction of the magnetic flux (this is the positive direction of the magnetic flux) while the other fingers indicating the positive direction of the current. As shown in Figure 2.24, the positive direction is the counterclockwise direction.

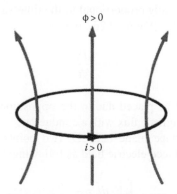

FIGURE 2.24   Linkage magnetic flux with a conductor ring.

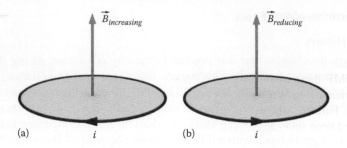

FIGURE 2.25   Induced current. (a) Increased magnetic flux; (b) Decreased magnetic flux.

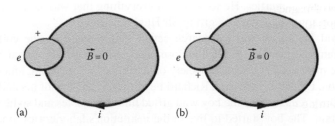

FIGURE 2.26   Equivalent circuit. (a) Increased magnetic flux; (b) Decreased magnetic flux.

Faraday observed that when the linkage magnetic flux with the ring varied with time, the continuous action resulted in an induced current. It was also verified that the direction of this current was always opposed to the magnetic flux variation, as shown in Figure 2.25 for two different situations.

The explanation for this induced current is due to the generating—what electric engineers prefer calling induction—of an electric field on the conductor created by the time-varying magnetic field.

The work done by a unit of electric charge—performed by this electric field—along the closed path is denominated electromotive force (EMF) and, according to Faraday's experiences, it is duly proportional to the time-varying rate of the linkage magnetic flux with conductor (Figure 2.26), that is,

$$e = -\frac{d\lambda}{dt}. \tag{2.39}$$

The negative sign shall be placed due to the opposition that this EMF exerts a variation of the linkage magnetic flux with a conductor.

So the calculus of voltage made between two points, EMF, in turn, can be expressed based on the induced electric field as following:

$$\oint_C \vec{E} \cdot d\vec{l} = -\frac{d\lambda}{dt}. \tag{2.40}$$

The meaning of the positive sign on the line integral at the left-hand side (2.40) is that the induced electric field exerts the role of an external agent which movement the electric charge, as we discussed when the concept of voltage between two points was presented.

The EMF dimension is *Volts*, which according to the International System of Units is identical to Wb/s.

At last, Faraday, who never had a formal education in childhood and adolescence, introduced time into electromagnetism, marking the birth of the field of electrical engineering and further discoveries.

2.3.5 MAGNETOMOTIVE FORCE

The concept of magnetomotive force (MMF) is an analogy to the EMF concept expressed in Equation 2.40 extracted from Ampère's circuital law, which is represented as

$$\oint_C \vec{H} \cdot d\vec{l} = i_t. \tag{2.41}$$

MMF is nothing more than the concatenated current with the closed path.

2.4 CONSTITUTIVE RELATIONS

The constitutive relations establish the physical properties of materials since they connect field vectors, which are the electric field $\vec{E}$ and magnetic field $\vec{B}$, associated with the sources $\vec{D}$, $\vec{H}$, and $\vec{J}$ to its effects.

For the magnetic field, we have already discussed the constitutive relations associated with it, which is

$$\vec{B} = \mu\vec{H}, \tag{2.42}$$

where μ is the magnetic permeability of the medium measured in the International System as (H/m). In the air (or vacuum) the value of magnetic permeability is é $\mu_0 = 4x\pi \times 10^{-7}$ H/m. This is a very low magnetic permeability, which shows a difficulty in establishing a magnetic field in air.

Frequently, the magnetic permeability of a medium is expressed as a multiple of the magnetic permeability of a vacuum through the relative permeability $\mu_r = \mu/\mu_0$.

Another form of expressing the same constitutive relation is through the following relation:

$$\vec{H} = \nu\vec{B}. \tag{2.43}$$

where ν is named magnetic reluctivity of the medium measured in m/H.

We also discussed the constitutive relations that connects the displacement vector $\vec{D}$ and the electric field $\vec{E}$, given by

$$\vec{D} = \varepsilon\vec{E}, \tag{2.44}$$

where ε is the electric permittivity of the medium measured in F/m. In the air (or vacuum) this permittivity value is $\varepsilon_0 = 10^{-9}/36\pi$ (F/m), which is the lower electric permittivity as possible.

Likewise, the relative electric permittivity of the medium is defined by the relation $\varepsilon_r = \varepsilon/\varepsilon_0$.

Finally, the constitutive relations relating the current density vector $\vec{J}$ with the electric field vector $\vec{E}$ is given by

$$\vec{J} = \sigma\vec{E}, \tag{2.45}$$

where σ is the conductivity of the medium expressed, in the International System, as S/m.

This relation is the expression of Ohm's law in terms of field vectors because the electric field is associated with the potential difference and the current density vector is associated with the electric current. Figure 2.27 shows an elementary current tube of cross section ΔS and length Δl.

The electric field and the current density vector in this elementary current tube can be expressed as $E = \Delta V/\Delta l$ and $J = \Delta i/\Delta S$, so that, on applying (2.45) we get

$$\frac{\Delta i}{\Delta S} = \sigma\frac{\Delta V}{\Delta l}.$$

Reorganizing the previous expression, we get

$$\Delta V = \frac{\Delta l}{\sigma\Delta S}\Delta i$$

or as $\Delta V = R\Delta i$, where $R = \Delta l/\sigma\Delta S$ is the resistance of the current tube.

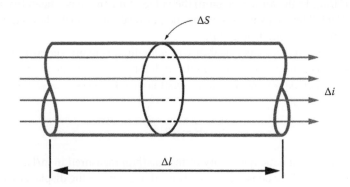

FIGURE 2.27 Elementary current tube.

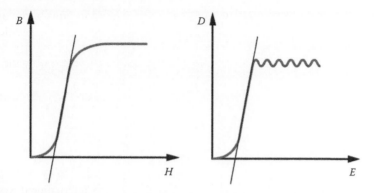

FIGURE 2.28   Constitutive relationship.

Physical properties μ, ε, and σ are parameters depending on a series of quantities. Generally, these properties depend on the field with which they are associated and on the temperature so that

$$\mu(B, T)$$

$$\varepsilon(E, T)$$

$$\sigma(E, T).$$

Excluding the conductivity σ, the influence of the temperature on both magnetic permeability μ and electric permittivity ε is only felt for very elevated values and normally small values are involved in engineering designs, for which—normally—it is considered $\mu = \mu(B)$ and $\varepsilon = \varepsilon(E)$; however, the temperature sensibly affects the conductivity.

Figure 2.28 shows the typical behavior of the constitutive relations of the main materials used in engineering.

Materials with variable properties according to the curves shown in Figure 2.28 are named nonlinear materials. Materials having constant physical properties for which their characteristic is a straight line, for example, of air (or vacuum), are named linear materials. Note that most of the materials used by engineering on their designs presents linearity for reduced values of fields, as shown in Figure 2.28.

Some materials also present behaviors depending on the field direction. Such materials, said anisotropic, present physical properties depending on the direction; Otherwise, the materials are called isotropic.

3 Maxwell's Equations

3.1 INTRODUCTION

James Clerk Maxwell was born in Edinburgh, Scotland, on June 13, 1831. He started showing his geniality soon in his adolescence. He published his first scientific paper when he was 14, where he described how to draw an ellipsis tying a line on two pins and dragging a pencil over a paper with a stretched-out line and the respective mathematic basis associated with the technique.

When he was 19, he entered University of Cambridge, and in the same year he published one of his renowned works about balance of elastic bodies. In 1854, he completed his undergraduate studies and started his graduate course, and the same year he was honored with the Smith award for his performance.

In 1855, he published *Experiments on Color as Perceived by the Eye* where he described its mathematical basis that led to the discovery of color television one century later. Maxwell created *dimensional analysis*, and he was the first to introduce *statistics* in a scientific study and the first to work on dynamic theory of gases.

When he was 27, Maxwell won the Adam award from the Royal Society (British scientific society and the first scientific society in history) for his work on the composition of Saturn rings that made them stable, preventing them to be attracted to the planet. His assertions were verified at the beginning of the 1980s with the help of photographs sent from the spacecraft *voyager*.

At the age of 30, he produced the first color photograph, as a result of the work done in 1855. In 1865, at the age of 34, He published his first scientific paper when he was 14, his work *A Dynamic Theory of the Electromagnetic Field* where he discussed the fundamentals of electromagnetism with a solid theory, which marked an era in history of humanity.

Maxwell performed several other relevant scientific works involving electromagnetism and the theory of gases and theory of control and compiled all the knowledge for *A Treatise on Electricity and Magnetism* published in 1873.

Maxwell's legacy is huge but not much acknowledged. Several inventions of his work are rarely connected to his name. Much is said about Hertzian waves in honor of Hertz when they should be Maxwellian waves because it was Maxwell who predicted them 30 years before. Much is said about radio being invented by Marconi but nowhere it is mentioned that he visited Maxwell in Cambridge before inventing it, and Marconi did not mention this event when he presented his work.

At the precocious end of his days, he passed away when he was 48. He found a research laboratory in Cambridge, which he named Cavendish Laboratory (in honor of Henry Cavendish). This laboratory has won, until now, 26 Nobel prizes, and some of the prize winners are J.J. Thomson, for the discovery of the electron, Rutherford, for the discovery of the model of the atom, and Watson and Crick, for the discovery of the DNA.

3.2 MAXWELL'S FIRST EQUATION

Maxwell admired Michael Faraday tremendously; he watched several of his presentations at the *Royal Institution of London,* one of the first organized research laboratories in history, and tried to become a member of the Royal Society of London with Faraday's nomination but he did not succeed. Faraday was very reserved and extremely religious, and he faced repeated aggressions from Humphry Davy, President of the Royal Institution, who was jealous of his lab assistant's success.

The death of Davy led Faraday to take over the presidency of that research laboratory. His work concerning magnetic induction was considered as the most important scientific event, which elevated the fundamental basis of modern science. The law of magnetic induction is the basis for Maxwell's first equation. Figure 3.1 shows the geometry used for the mathematical description of the magnetic induction phenomenon.

In Equation 3.1, Ø indicates a time-dependent linkage magnetic flux with the closed path C being arbitrarily oriented.

On August 29, 1831, Faraday established that when a closed path subjected to a time-varying leakage magnetic flux appears, an induced electromotive force (EMF) is given by

$$e = -d\emptyset/dt. \tag{3.1}$$

This EMF can be expressed in terms of electric field vector through the work performed by this field for dragging a unit electric charge in the closed path C, that is,

$$e = \oint_C \vec{E} \cdot \vec{dl}. \tag{3.2}$$

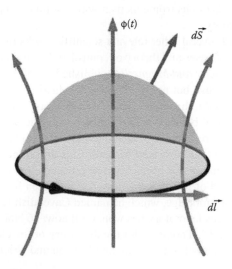

FIGURE 3.1 Law of magnetic induction.

The linkage magnetic flux in the closed path C can also be expressed in terms of magnetic field vector as following:

$$\phi = \int_S \vec{B} \cdot d\vec{S}. \tag{3.3}$$

Replacing this quantity in Equation 3.1 results in

$$\oint_C \vec{E} \cdot d\vec{l} = -\frac{d}{dt} \int_S \vec{B} \cdot d\vec{S}.$$

As S and t are independent state variables, that is, the surface neither expands nor contracts with time, we can write

$$\oint_C \vec{E} \cdot d\vec{l} = -\int_S \frac{\partial \vec{B}}{\partial t} \cdot d\vec{S}. \tag{3.4}$$

This is the *integral* form of Maxwell's first equation.

Everyday Maxwell's first equation is frequently used in electrical engineering by students without realizing that they are manipulating it.

If $\vec{B}$ is constant in time, the induced EMF is null because the rate of variation of linkage magnetic flux in time is null. Thus, in this case we can write

$$\oint_C \vec{E} \cdot d\vec{l} = 0. \tag{3.5}$$

This result implies the following:

1. *The voltage—in stationary regime ($\partial/\partial t = 0$)—between any two points does not depend on the path used for the integration.*

   For demonstrating this, two points A and B are considered immersed in a region where a magnetic field constant with time is present, as shown in Figure 3.2.

   V_1 is the voltage between A and B evaluated by the path 1, that is,

$$V_1 = -\int_{A1B} \vec{E} \cdot d\vec{l}.$$

   V_2 is the voltage between A and B evaluated by the path 2, that is,

$$V_2 = -\int_{A2B} \vec{E} \cdot d\vec{l}.$$

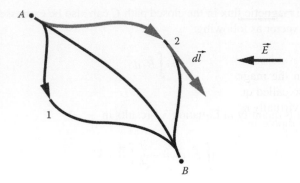

FIGURE 3.2   Voltage.

As EMF induced in the closed path constituted by paths 1 and 2 is null because there is no time variation of the magnetic field; thus, we can write

$$\oint_C \vec{E} \cdot d\vec{l} = -V_1 + V_2 = 0. \tag{3.6}$$

Consequently, $V_1 = V_2$.

Any vector field with this characteristic is said to be *conservative*, which evidences that the work performed by this field in a closed path is always null.

2. *The sum of voltage drops in a mesh of an electric circuit is null: Kirchhoff's second law.*

Figure 3.3a shows a single mesh electric circuit constituted by four dipoles and $\vec{B}$ as time-varying magnetic field, generated or not by its own current, which produces a leakage magnetic flux in the circuit's mesh. The equivalent electric circuit shown in Figure 3.3b, without the

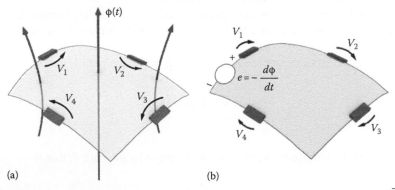

(a)   (b)

FIGURE 3.3   Induced EMF effect. (a) Real electric circuit $V_1 + V_2 + V_3 + V_4 = -\int_S \dfrac{\partial \vec{B}}{\partial t} \cdot d\vec{S}$. (b) Equivalent electric circuit $V_1 + V_2 + V_3 + V_4 = -e$.

leakage magnetic flux in its closed path, reproduces same conditions as the previous one.

Figure 3.3b includes an additional electric voltage source, whose EMF is the one evaluated by Faraday's law applied to the circuit's sides.

When the magnetic field is constant or with a very slow time variation (also called quasi-static magnetic field), we can consider $dl/dt \approx 0$; the EMF is virtually null and the result obtained from the sum of voltage drops on each dipole in a mesh is null, which coincides with the application of Kirchhoff's second law. Maxwell's first equation can also be written in a differential form, using field operators.

3.2.1 THE CURL OF A VECTOR FIELD

We are ready now for introducing the concept of field operators. As we need the curl for expressing Maxwell's first equation in a differential form, let us discuss now the meaning of this field operator. For such, let us solve a simple problem of mechanics that consists in calculating the work performed by a force in dragging a body along a closed path, as shown in Figure 3.4 [40,41].

The work performed in dragging a body along a closed path is given by

$$\tau = \oint_C \vec{F} \cdot \vec{dl}. \tag{3.7}$$

If the evaluation of this integral is difficult, a numerical alternative can be used for solving the problem. One of the possibilities is subdividing the surface of the closed path into a very large number of small elements with small dimensions to consider $\vec{F}$ constant at the sides of each element, as shown in Figure 3.5.

After discretized the surface the work performed in moving the body along each elementary closed path, of same shape is calculated than the prior, therefore, due to its elementary dimensions, the evaluation of this work become simple, as expression (3.8) shows

$$\Delta\tau = \oint_{\Delta C} \vec{F} \cdot \vec{dl} = \sum_{i=1}^{4} \vec{F_i} \cdot \vec{\Delta l_i}. \tag{3.8}$$

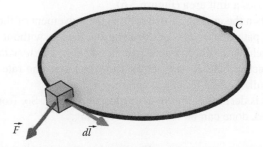

FIGURE 3.4 Work performed in dragging a body.

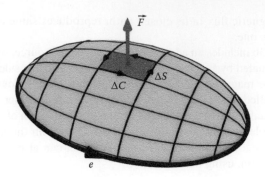

FIGURE 3.5 Subdivision of the closed path.

The sum of the values for work calculated over each elementary closed path results, approximately, in the total amount of work required in dragging the body along the original closed path. Note that during calculating the total amount of work over the elementary closed path, the work of the adjacent sides are canceled. Thus, we can write

$$\tau = \sum_{i=1}^{n} \Delta\tau = \sum_{i=1}^{n} \oint_{\Delta C} \vec{F} \cdot d\vec{l}. \tag{3.9}$$

We need to understand the curl of $\vec{F}$. For such, we isolate an element ΔC on the internal surface of the closed path. It is defined as the curl of the force $\vec{F}$, designed by curl $\vec{F}$, or $\nabla \times \vec{F}$, in a given point, with the following characteristics:

$$\text{Amplitude:} \left| \nabla \times \vec{F} \right| = \lim_{\Delta s \to 0} \frac{\Delta\tau}{\Delta S}, \tag{3.10}$$

Direction: Normal to surface,
Sense: The positive sense is pointed by thumb while other fingers point toward the given orientation of the closed path.

Physically, the curl $\vec{F}$ represents the work performed in moving the body along a closed path that closes a unit area (Figure 3.6).

Observe that when the curl of a force is null, the movement of the body along the elementary closed path occurs without dissipation, that is, without friction. This is the reason why a field with this characteristic is said to be *conservative*. This is what occurs with the electric field $\vec{E}$ on electrostatic when the time rate variation of the magnetic field is null.

Stokes theorem is derived from the definition of the curl. So, from Equation 3.10 the elementary work done can be written as following:

$$\Delta\tau = \left| \nabla \times \vec{F} \right| \cdot \Delta S. \tag{3.11}$$

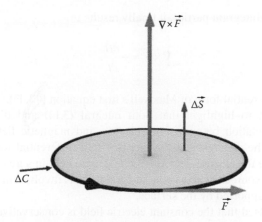

FIGURE 3.6 Curl $\vec{F}$ at a given point.

The sum of all elementary works results in the total work performed along the closed path C, for that

$$\tau = \sum_{i=1}^{n} \Delta\tau = \sum_{i=1}^{n} \left[\nabla \times \vec{F}\right]_i \cdot \Delta\vec{S}_i. \tag{3.12}$$

When $n \to \infty$, we obtain

$$\oint_C \vec{F} \cdot d\vec{l} = \int_S \nabla \times \vec{F} \cdot d\vec{S}. \tag{3.13}$$

This is the expression of Stokes theorem, which will be discussed later.
Coming to Maxwell's first equation, which we rewrite by convenience, we have

$$\oint_C \vec{E} \cdot d\vec{l} = -\int_S \frac{\partial \vec{B}}{\partial t} \cdot d\vec{S}. \tag{3.14}$$

Applying Stokes theorem to the left-hand side of this equation, we get

$$\oint_C \vec{E} \cdot d\vec{l} = \int_S \nabla \times \vec{E} \cdot d\vec{S}. \tag{3.15}$$

Replacing its result in Equation 3.14 results in

$$\int_S \nabla \times \vec{E} \cdot d\vec{S} = -\int_S \frac{\partial \vec{B}}{\partial t} \cdot d\vec{S}. \tag{3.16}$$

Identifying the integrant parties, it finally results in

$$\nabla \times \vec{E} = -\frac{\partial \vec{B}}{\partial t}. \tag{3.17}$$

This is the differential form of Maxwell's first equation [42,43].

It is important to highlight that both integral (3.14) and differential forms (3.17) express a relation between both electric and magnetic fields as Faraday's law establishes. The difference is the fact that in a differential form, a relation is expressed between the electric field and the magnetic field in a given point of space, independent of its sources, whereas the integral form is involved in a region defined by a closed path supported by the surface.

We have discussed that the constant electric field is conservative because in this case it is verified that $\nabla \times \vec{E} = 0$. It implies uniqueness voltage between two points because this value does not depend on the integration path of the electric field used for obtaining it. However, in time-varying fields, the electric field is not conservative and becomes dissipative because $\nabla \times \vec{E} \neq 0$.

What difference does it produce on the voltage? The answer is easy that there is not only one value for the voltage between two points. In that case, the voltage depends on the path chosen for integration. Let us discuss this question to analyze a simple problem of an electric circuit with a single rectangular mesh constituted of two different resistors R_1 and R_2, as shown in Figure 3.7.

Let us consider that this mesh is subjected to a time-varying leakage magnetic flux, as indicated in the previous figure. By Faraday's law, a current is induced in this mesh, which opposes the linkage magnetic flux variation. The question presented is the following: What is the voltage between points A and B of the circuit shown in Figure 3.7? This is a typical case of a voltage between two points not being the same, because its value depends on how its measurement is performed.

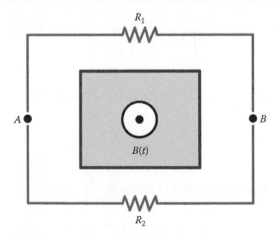

FIGURE 3.7 Electric circuit.

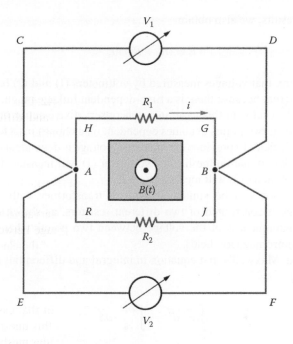

FIGURE 3.8 Measurement of the voltage.

Figure 3.8 shows both measurement possibilities of this voltage using a voltmeter.

Starting with measurement alternative using voltmeter (1), located at the upper side of Figure 3.8, both meshes are identified as follows:

$$\text{Mesh } \alpha: \text{ACDBGHA}$$

$$\text{Mesh } \beta: \text{AKJBDCA}$$

When applying Kirchhoff's second law to the mesh α, we obtain $V_1 = R_1 \cdot i$. Observe that there is no magnetic flux enclosed with this mesh.

When applying this law to mesh β, we should consider the effect of the concatenated magnetic flux on this mesh, which is responsible for the induction of an induced EMF represented as E. Thus, Kirchhoff's second law applied to the mesh β results in $V_1 = E - R_2 \cdot i$. It is clear that, from these results, we obtain

$$E = (R_1 + R_2) \cdot i.$$

Starting with measurement alternative using voltmeter (1), located at the lower side of Figure 3.8, results in the following measurements:

$$\text{Mesh } \delta: \text{AEFBJKA} - V_2 = R_2 \cdot i$$

$$\text{Mesh } \gamma: \text{AEFBGHA} - V_2 = E - R_1 \cdot i$$

From these results, we also obtain

$$E = (R_1 + R_2) \cdot i.$$

Note, therefore, that voltages measured by voltmeters (1) and (2) between points A and B are different because there is a time-dependent linkage magnetic flux in the electric circuit. It results, in this case, in a nonconservative electric field, for which the voltage between two points becomes dependent on a chosen path for the integration of the electric field. In previous example, this voltage is dependent on the routing of voltmeter cables so that on mesh α the voltmeter (1) will measure $V_1 = R_1 \cdot i$ and on mesh δ the voltmeter (2) will measure $V_2 = R_2 \cdot i$.

This phenomenon can be simulated using a transformer with its secondary replaced by a mesh constituted of two different resistors, as shown in Figure 3.7, indicating a nonuniqueness of the voltage between two points when dealing with time-varying electromagnetic fields.

Summarizing, Maxwell's first equation in integral and differential forms is written as following:

$$\oint_C \vec{E} \cdot d\vec{l} = -\int_S \frac{\partial \vec{B}}{\partial t} \cdot d\vec{S}. \tag{3.18}$$

$$\nabla \times \vec{E} = -\frac{\partial \vec{B}}{\partial t}. \tag{3.19}$$

3.3   MAXWELL'S SECOND EQUATION

3.3.1   THE DISPLACEMENT CURRENT

Maxwell studied all the pivotal experiments of both electricity and magnetism performed by his antecessors, aiming to consolidate a theory that contemplates all the relations between electromagnetic fields. To achieve that objective, it was necessary to check the balance of charges since charge conservation was an established fundamental principle and universally accepted. This checking did not match when Faraday's law was considered. The difference needed to be "adjusted" for assuring the conservation law of electric charges. For solving this problem, Maxwell started from the expression of the current crossing a surface in terms of the current density vector (Figure 3.9):

$$i = \int_S \vec{J} \cdot d\vec{S}. \tag{3.20}$$

Next, he agreed to the hypothesis that the surface is closed. It is well established that the total electric current crossing a closed surface is null; as a profit of this thought, Kirchhoff's first law was established (nodes law) stating that the total amount of electric current leaving a circuit's node is null. This assertion is only true

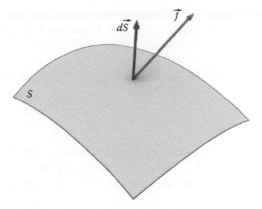

FIGURE 3.9   Current over a surface.

for phenomena of "stationary regime" that are constant in time where $d/dt = 0$; this was the predominant knowledge of the scientific community prior to Faraday's work.

Thus, admitting the closed surface S, we write (Figure 3.10)

$$\oint_{\Sigma} \vec{J} \cdot d\vec{S} = i, \tag{3.21}$$

where Σ denotes a closed surface with elementary surface vector $d\vec{S}$ pointed outside.

In Equation 3.21 the current i will only be different from zero if a variation occurs in the quantity of electric charges inside the surface (it can happen, as we will see). For example, if a positive current leaves the closed surface, it means

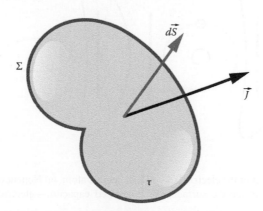

FIGURE 3.10   Current "leaving" a closed surface.

that positive charges leave the interior of the surface and the total amount of internal electric charges decreases, for which Equation 3.21 can be rewritten as following:

$$\oint_{\Sigma} \vec{J} \cdot d\vec{S} = -\frac{dQ}{dt}. \tag{3.22}$$

Q is the total amount of electric charges inside the surface. The negative sign placed on the right side of Equation 3.22 indicates $i > 0$ leaving the surface, which implies negative time-varying rate in the total amount of electric charges inside the closed surface.

Capacitor's Case

This phenomenon occurs, for example, during the charging of a capacitor. For understanding it, consider a closed surface involving only one of the capacitor's parallel plates, as shown in Figure 3.11. With the nonenergized capacitor (Figure 3.11a), the quantity of electric charges internal to the surface is null. Once energized, the capacitor is charged for which the plate involved in the closed surface receives the quantity of electric charges (Figure 3.11b) different from zero. Thus, it is easy to understand that during the time interval corresponding the charging of the capacitor, electric charges only enter the surface violating, in principle, Kirchhoff's law of currents, as we know it.

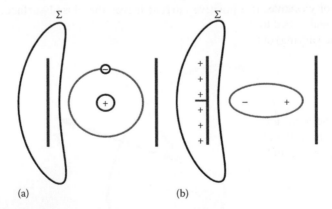

(a)   (b)

FIGURE 3.11   Effect of the electric field on a dielectric atom. (a) Nonenergized capacitor—electric charge null inside the surface. (b) Energized capacitor—electric charge non-null inside the surface.

On the other hand, the quantity of internal charges to the surface can be expressed based on the displacement vector $\vec{D}$, from Gauss' law, as following:

$$\oint_{\Sigma} \vec{D} \cdot d\vec{S} = Q.$$

Replacing this result in Equation 3.22, we obtain

$$\oint_{\Sigma} \vec{J} \cdot d\vec{S} = -\frac{d}{dt} \left[\oint_{\Sigma} \vec{D} \cdot d\vec{S} \right]. \tag{3.23}$$

As the surface Σ is nondeformable, a time variation derivative can be introduced in the integral of the right-hand side. Besides, the integration variables of both sides are the same, for that we can join their integrant parts in a single integral as following:

$$\oint_{\Sigma} \left(\vec{J} + \frac{\partial \vec{D}}{\partial t} \right) \cdot d\vec{S} = 0. \tag{3.24}$$

It is convenient now to reflect a little bit on the result obtained. Since the expression is dimensionally correct, the term $\partial \vec{D} / \partial t$ should have the dimension of the current density. In fact, as $\vec{D}$ has the dimension C/m² and time s, the quotient between these two measuring units is A/m², which is the dimension of the current density.

The current density vector $\vec{J}$ is associated to the electric field vector through Ohm's law ($\vec{J} = \sigma\vec{E}$). This current density corresponds to the one associated to the movement of electrons located in the last shell of atoms in conducting materials, which are weakly tied to the nucleus and for which, independent in behavior (if constant or time varying) of the electric field, a type of electric current is generated. This kind of electric current is termed "conduction current."

The current density $\vec{J}_D = \partial \vec{D} / \partial t$ is associated to the displacement (this is the reason for the name displacement vector) of the electronic cloud of atoms when subjected to a time-dependent electric field. Let us analyze the following situation: imagine a capacitor with an ideal dielectric. While the electric field of the capacitor is null, the electronic cloud can be imagined as a sphere for which the centers of positive and negative charges are coincident, as shown in Figure 3.11a. Imposing an electric field on the dielectric through the application of a voltage between the plaques of the capacitor, a deformation occurs in the electronic cloud (the positive plate attracts the electronic cloud and the negative plate attracts the nucleus), which can be associated to an ellipsoid, promoting a noncoincidence among the charge centers.

The displacement of the electric charge from the condition shown in Figure 3.11a to the condition shown in the Figure 3.11b corresponds to an electric current without the migration of electrons from one atom to another. This type of electric current is called displacement current. As a conclusion the displacement current is non-null only time-dependent electric field. In a constant field, the electronic cloud initially deforms becoming capacitor's current charge, which once charged, the current becomes null again because the cloud, although deformed, becomes static.

This type of current is also presented in conductors subjected to a time-dependent electric field due to the electrons of interior layers, which are strongly attracted to the nucleus. In this case, the electrons of the last layer move from one atom to another producing a conduction current, and the electrons of interior layers vibrate with varying field producing a displacement current.

In good conducting materials, the conduction current is very superior to the displacement current when subjected to a time-varying field, that is, $\vec{J} \gg \partial \vec{D}/\partial t$, while in case of good dielectrics the inverse occurs [44,45].

3.3.2 TOTAL CURRENT CROSSING A SURFACE

After the identification of a displacement current, Maxwell proposed attributing to them same properties as attributed to the conduction current, particularly in what concerns the production of magnetic fields.

Thus, if a given medium is subjected to a time-dependent electric field, this medium will host both types of electric current: the conduction current and the displacement current.

So, given an open surface immersed in a time-dependent electric field, as shown in Figure 3.12, the total current crossing this surface is given by

$$i_t = \int_S \left(\vec{J} + \frac{\partial \vec{D}}{\partial t} \right) \cdot d\vec{S}. \tag{3.25}$$

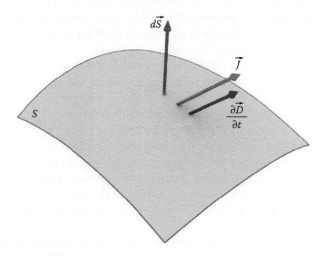

FIGURE 3.12   Total current crossing an open surface.

As Maxwell extended the properties of the conduction current to the displacement current, we need to rewrite Ampère's circuital law taking into account both types of current.

For a given closed path, supporting a surface in a conducting medium, it can involve the two types of current and the current linkaged from this closed path can be evaluated from Equation 3.25 for which "Ampère's circuital law" is written as following:

$$\oint_C \vec{H} \cdot d\vec{l} = \int_S \left(\vec{J} + \frac{\partial \vec{D}}{\partial t} \right) \cdot d\vec{S}. \tag{3.26}$$

This is the integral form of Maxwell's second equation, which generalized "Ampère's circuital law" considering the displacement current as the total current linkaged from closed path C.

Applying Stokes theorem to the left-hand side of Equation 3.26, we obtain

$$\oint_C \vec{H} \cdot d\vec{l} = \int_S \nabla \times \vec{H} \cdot d\vec{S}. \tag{3.27}$$

Replacing this result again to the left-hand side of Equation 3.26 and identifying the integrant parties, it results in

$$\nabla \times \vec{H} = \vec{J} + \frac{\partial \vec{D}}{\partial t}. \tag{3.28}$$

This is the differential form of Maxwell's second equation.

Summarizing, Maxwell's second equation in integral and differential forms is written as following:

$$\oint_C \vec{H} \cdot d\vec{l} = \int_S \left(\vec{J} + \frac{\partial \vec{D}}{\partial t} \right) \cdot d\vec{S}. \tag{3.29}$$

$$\nabla \times \vec{H} = \vec{J} + \frac{\partial \vec{D}}{\partial t}. \tag{3.30}$$

Observe that the magnetic field is not conservative because $\nabla \times \vec{H} \neq 0$, and also observe that imposing the constant field, that is, $\partial/\partial t = 0$ it reproduces the original for the "Ampère's circuital law."

3.4 MAXWELL'S THIRD EQUATION

Maxwell's third equation translates the behavior of the magnetic flux density field lines. From experimental observations it is verified that these lines are closed. Vector fields with these characteristics are the ones that do not have punctual sources. It is a consequence of

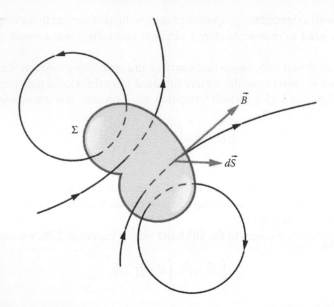

FIGURE 3.13 Magnetic field lines.

the fact that isolated magnetic charges do not exist, but that does not occur in an electric field that has two types of electric charges, positive and negative, as sources that are presented in nature separately. For the magnetic flux density field, its analogs are the North and South poles, which are not presented separately in nature, that is, the existence of a North pole is always associated to the existence of a South pole.

Figure 3.13 shows a closed surface Σ immersed in a region where a magnetic field is present.

As its lines are closed, the total amount of lines that enters the surface is exactly equal to the total amount of lines that leaves it. For this reason, the magnetic flux over a closed surface Σ is null.

Translating mathematically this experimental observation, it results in

$$\oint_{\Sigma} \vec{B} \cdot d\vec{S} = 0. \tag{3.31}$$

This is the expression of Maxwell's third equation in an integral form.

3.4.1 DIVERGENCE OF A VECTOR FIELD

The expression of Maxwell's third equation in a differential form uses the divergence concept of a vector field. This concept is very simple, and it is associated to the type of source generating it.

Let us analyze the problem of hydraulic nature for its easy understanding. Suppose a tank, totally closed, is filled with gas, without considering the shape, and imagine that the shape of this tank is rather unusual. It is evident that under these conditions,

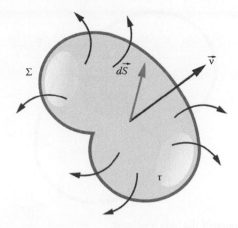

FIGURE 3.14 Gas flow calculus.

the gas flux through the tank's surface is null. Imagine a large quantity of small holes are made throughout the tank's surface so that gas flow is initiated generalized on the tank's surface (Figure 3.14).

For evaluating the gas flow through the tank's surface, we should calculate the speed vector flux of the gas over the closed surface, that is,

$$Q(m^3) = \oint_\Sigma \vec{v} \cdot d\vec{S}. \tag{3.32}$$

If the analytic evaluation of this integral is difficult, an approximated method could be applied by subdividing the tank's volume (τ) into a very large number of small volumes ($\Delta\tau$), as shown in Figure 3.15.

If the volumes obtained from this subdivision are sufficiently small, the calculus of the integral (3.32) becomes easier because the speed can be considered constant on each side of the elementary volume. Therefore, the flux on the elementary volume is given by

$$\Delta Q(m^3) = \oint_{\Delta\Sigma} \vec{v} \cdot d\vec{S} = \sum_{i=1}^{6} \vec{v}_i \cdot \Delta\vec{S}_i. \tag{3.33}$$

Once the elementary flux of all the volume elements is evaluated, the total flux is obtained from the sum of all elementary flux.

The divergence concept of the speed vector is directly associated to the elementary flux, that is, the divergence of speed expressed as $\nabla \cdot \vec{v}$ or $div\ \vec{v}$ in a point is the elementary flux of the speed vector by volume unit at this point or

$$\nabla \cdot \vec{v} = \lim_{\Delta\tau \to 0} \frac{\oint_{\Delta\Sigma} \vec{v} \cdot d\vec{S}}{\Delta\tau}. \tag{3.34}$$

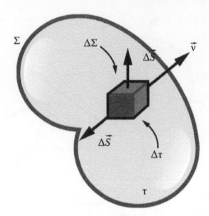

FIGURE 3.15   Subdivision of the tank's volume.

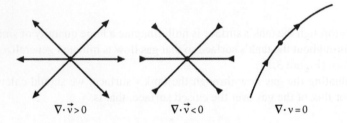

FIGURE 3.16   Divergence of a vector field.

Thus, if divergence of the speed vector in a point is positive, it means that the gas is escaping from an elementary volume at this point and it is said, in this case, that at this point there is a "source" of field. In case the divergence is negative, it means that the gas is reaching an elementary volume at the referred point and it is said that at this point there is a "sink" of field. Finally, if the divergence is null, the gas volume is maintained constant. Figure 3.16 illustrates these three conditions.

3.4.2   Gauss's or Divergence One's Theorem

Gauss's theorem is obtained from relation (3.34) as follows [46]:

$$\nabla \cdot \vec{v} \Delta \tau = \oint_{\Delta \Sigma} \vec{v} \cdot d\vec{S}. \tag{3.35}$$

Applying this relation to all the n elementary volumes obtained from subdivided domain, while considering that $n \rightarrow \infty$ and performing the sum on both sides, the prior expression is written as follows:

$$\int_{\tau} \nabla \cdot \vec{v} \cdot d\tau = \oint_{S} \vec{v} \cdot d\vec{S}. \tag{3.36}$$

The result obtained in Equation 3.36 is the expression of Gauss's theorem, also named as divergence theorem, which will be detailed further.

Applying Gauss's theorem to Maxwell's third equation in an integral form, we obtain

$$\int_{\tau} \nabla \cdot \vec{B} \, d\tau = 0.$$

It results that the integrant shall be null, that is,

$$\nabla \cdot \vec{B} = 0. \tag{3.37}$$

This is the expression of Maxwell's third equation in a differential form, which evidences the fact about the nonexistence of punctual sources of magnetic field.

Summarizing, Maxwell's third equation in both integral and differential forms is written as following:

$$\oint_{\Sigma} \vec{B} \cdot d\vec{S} = 0. \tag{3.38}$$

$$\nabla \cdot \vec{B} = 0. \tag{3.39}$$

3.5 MAXWELL'S FOURTH EQUATION

Maxwell's fourth equation is derived from electrostatic Gauss's law, which establishes that the displacement vector flux over a closed surface is equal to its internal charge (Figure 3.17), that is,

$$\oint_{\Sigma} \vec{D} \cdot d\vec{S} = Q. \tag{3.40}$$

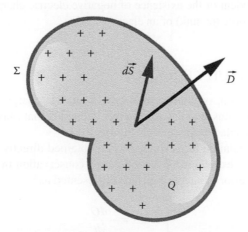

FIGURE 3.17 Maxwell's fourth equation.

As the internal charge on the surface can be expressed based on the volumetric density of charges over the volume, so we can write

$$\oint_{\Sigma} \vec{D} \cdot d\vec{S} = \int_{\tau} \rho_v \, d\tau. \tag{3.41}$$

The expression (3.41) is the integral form of Maxwell's fourth equation. The differential form is obtained applying Gauss's theorem to the left-hand side of the equation as follows:

$$\int_{\tau} \nabla \cdot \vec{D} \, d\tau = \int_{\tau} \rho_v \, d\tau$$

As the integration variables of both sides are equal, we can identify the integrands, for that

$$\nabla \cdot \vec{D} = \rho_v. \tag{3.42}$$

This is the differential form of Maxwell's fourth equation.

Summarizing, Maxwell's fourth equation in both integral and differential forms is written as following:

$$\oint_{\Sigma} \vec{D} \cdot d\vec{S} = \int_{\tau} \rho_v d\tau. \tag{3.43}$$

$$\nabla \cdot \vec{D} = \rho_v, \tag{3.44}$$

which evidences the fact about the possibility for isolated existence of positive electric charges independent of the existence of negative electric charges, or vice versa. Such charges are source (or sink) of an electric field.

3.6 THE CONTINUITY EQUATION

The continuity equation, or conservation law of electric charges, can be deduced from Maxwell's equations, and it was from this deduction that Maxwell felt the need of introducing the displacement current (Figure 3.18).

However, its formulation turns simpler when obtained directly from the result in Equation 3.22, which expresses the law of charge conservation in terms of quantity of charges in the interior of a closed surface represented as

$$\oint_{\Sigma} \vec{J} \cdot d\vec{S} = -\frac{dQ}{dt}. \tag{3.45}$$

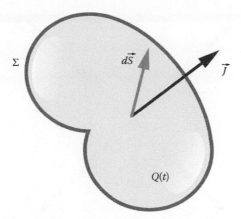

FIGURE 3.18 Continuity equation.

As the internal charge on the surface can be expressed in terms of volumetric density of electric charges, we can write

$$\oint_{\Sigma} \vec{J} \cdot d\vec{S} = -\frac{d}{dt}\left(\int_{\tau} \rho_v \, d\tau\right). \tag{3.46}$$

As the integration variables dt and $d\tau$ are independent variables, because it is admitted that the involved geometries are time-unvarying geometries, the derivative related to time can be introduced in the integral's interior resulting in

$$\oint_{\Sigma} \vec{J} \cdot d\vec{S} = -\int_{\tau} \frac{\partial \rho_v}{\partial t} \, d\tau. \tag{3.47}$$

This is the integral form of the continuity equation establishing the fact that if there is an electric current leaving a closed surface, there is a decrease in the total amount of electric charges internal to the closed volume of this surface.

The differential form of this integral equation is obtained applying Gauss's theorem to the left-hand side of Equation 3.47 as following:

$$\int_{\tau} \nabla \cdot \vec{J} \, d\tau = -\int_{\tau} \frac{\partial \rho_v}{\partial t} d\tau. \tag{3.48}$$

As the integration variables of both sides are equal, we can identify the integrands resulting in

$$\nabla \cdot \vec{J} = -\frac{\partial \rho_v}{\partial t}. \tag{3.49}$$

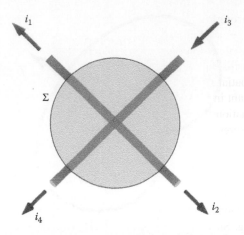

FIGURE 3.19 Continuity equation in stationary regime.

This is the differential form of the continuity equation. Note that in the case of constant fields—which implies $\partial/\partial t = 0$—the current density vector flux on a closed surface, or more precisely, the total current leaving a closed surface is null, as Kirchhoff's law of nodes, similar to the current distribution shown in Figure 3.19.

The continuity equation applied to the closed surface that involves a node of the circuit in DC current supplies

$$\int_{\Sigma} \vec{J} \cdot d\vec{S} = i_1 + i_2 - i_3 + i_4 = 0.$$

Summarizing, the continuity equation in both integral and differential forms is written as following:

$$\oint_{\Sigma} \vec{J} \cdot d\vec{S} = -\int_{\tau} \frac{\partial \rho_v}{\partial t} \, d\tau. \tag{3.50}$$

$$\nabla \cdot \vec{J} = -\frac{\partial \rho_v}{\partial t}, \tag{3.51}$$

which evidences the fact about the possibility for the existence of punctual sources (or sink) of the current density vector in case of a punctual variation of the volumetric density of charges.

3.7   WHERE ARE WE?

We just finished discussing Maxwell's equations, which are the mathematical representations of electricity's experimental physics laws discovered in the 1860s by different researchers. The highlight on these equations, which had not been discovered through experiments, was the introduction of the "displacement current" concept of Ampère's law.

Maxwell also contributed to mathematical formalism through the presentation of differential forms of his equations. These differential forms of Maxwell's equation represent the first spatial relations between both electrical and magnetic fields that became very important in all kinds of field studies.

Thus, its first equation is Faraday's law of magnetic induction, which, in stationary regime, is the expression of Kirchhoff's law of meshes; Ampère's circuital law, as mentioned earlier, is based on Maxwell's second equation, which contemplates the effects of displacement current. Maxwell's third equation highlights the evidence of the nonexistence of punctual sources of magnetic field, and finally Maxwell's fourth equation is derived from electrostatic Gauss's law. These equations describe any electromagnetic phenomenon and, in their essence, summarize all the knowledge about electrical engineering. Since the establishment of these equations, the investment of scientific community was toward the direction of developing mathematical techniques for the solution of these equations to solve practical problems in electrical engineering.

Table 3.1 summarizes results obtained till now, which are very useful for a better understanding of the geometry involved in each equations.

It occurs, however, that Maxwell discovered a pearl when manipulating them, which revolutionized concepts not known before about the behavior of electromagnetic fields, which will be discussed in the next chapter.

TABLE 3.1

Maxwell's Equations and Continuity Equation

Equation	Integral Form	Differential Form	Geometry
I	$\oint_c \vec{E} \cdot \vec{dl} = -\int_s \frac{\partial \vec{B}}{\partial t} \cdot \vec{dS}$	$\nabla \times \vec{E} = -\frac{\partial \vec{B}}{\partial t}$	
II	$\oint_c \vec{H} \cdot \vec{dl} = \int_s \left(\vec{J} + \frac{\partial \vec{D}}{\partial t} \right) \cdot \vec{dS}$	$\nabla \times \vec{H} = \vec{J} + \frac{\partial \vec{D}}{\partial t}$	
III	$\oint_\Sigma \vec{B} \cdot \vec{dS} = 0$	$\nabla \cdot \vec{B} = 0$	
IV	$\oint_\Sigma \vec{D} \cdot \vec{dS} = \int_\tau \rho_v \, d\tau$	$\nabla \cdot \vec{D} = \rho_v$	
Continuity equation	$\oint_\Sigma \vec{J} \cdot \vec{dS} = -\int_\tau \frac{\partial \rho_v}{\partial t} d\tau$	$\nabla \cdot \vec{J} = -\frac{\partial \rho_v}{\partial t}$	

3.8 A LITTLE BIT OF HISTORY

An interesting exercise is to imagine how Maxwell's equations were written during different ages of electromagnetism history.

3.8.1 STATIONARY REGIME

Let us imagine, however, Maxwell's equations written in eighteenth century during the time of Benjamin Franklin, Henry Cavendish, Charles Coulomb, Gauss, Weber, Volta, etc. At that time, field of electricity was detached from magnetism, there was no knowledge about magnetic field produced by an electric current, and magnetism was only produced by permanent magnets and by terrestrial magnetism.

In that century, the electric field and magnetic field was time constant (i.e., $\partial/\partial t = 0$), and it was unknown that $\vec{J}$ would produce magnetic field. In this case, Maxwell's equations can be divided into two groups, as shown in Table 3.2.

The first group presents electrostatic equations—the electricity from that time—whose science was analyzed experimentally for the actions between stationary electric charges. The second group presents magnetism equations that describe the magnetic field behavior not produced by electric currents, such as the ones produced by permanent magnets and by terrestrial magnetism.

In the beginning of nineteenth century, more precisely in 1821, the work of Hans Christian Oersted arose, which showed evidences of the electric current issued from Volta's battery (DC electric current) producing magnetic field—improving the equations of magnetism—due to which magnetism equations took into account the current density vector $(\vec{J})$, as shown in Table 3.3.

TABLE 3.2
Maxwell's Equations in Eighteenth Century

Equation	Integral Form	Differential Form
Electricity (electrostatics)		
I	$\oint_C \vec{E} \cdot d\vec{l} = 0$	$\nabla \times \vec{E} = 0$
IV	$\oint_\Sigma \vec{D} \cdot d\vec{S} = \int_\tau \rho_v \, d\tau$	$\nabla \cdot \vec{D} = \rho_v$
Magnetism		
II	$\oint_C \vec{H} \cdot d\vec{l} = 0$	$\nabla \times \vec{H} = 0$
III	$\oint_\Sigma \vec{B} \cdot d\vec{S} = 0$	$\nabla \cdot \vec{B} = 0$

TABLE 3.3
Magnetism Equations after 1821 and before 1831

Equation	Integral Form	Differential Form
II	$\oint_C \vec{H} \cdot d\vec{l} = \int_S \vec{J} \cdot d\vec{S}$	$\nabla \times \vec{H} = \vec{J}$
III	$\oint_\Sigma \vec{B} \cdot d\vec{S} = 0$	$\nabla \cdot \vec{B} = 0$

3.8.2 QUASI-STATIC REGIME

After Faraday's work, which showed that the time-dependent magnetic field could also produce an electric field, Maxwell's equations were written, in 1831, as shown in Table 3.4. Observe that at that time, the *displacement current* was not known.

Faraday can take full credit for the introduction of the variable (t)—the time—at Maxwell's equations. Table 3.4 expresses Maxwell's equations suitable for modeling quasi-static electromagnetic field problem in electrical engineering as the low-frequency electromagnetic phenomenon, as is the case in electromechanical devices like motors, generators, transformers, etc. In this type of study, the displacement current is very smaller than the conduction current, that is, $J \gg \partial D/\partial t$, so that we can neglect it in view of this evidence.

3.8.3 COMPLEX REPRESENTATION OF SINUSOIDAL QUANTITIES

The electrical equipment, even though they are for both industrial and residential application, normally include electric motors, generators, transformers, and illumination fed by electrical network from the power electric company. This electrical network, in turn, supplies the electric power to consumers under a form of sinusoidal time-dependent

TABLE 3.4
Maxwell's Equations after 1831 and before 1867

Equation	Integral Form	Differential Form
I	$\oint_C \vec{E} \cdot d\vec{l} = -\int_S \frac{\partial \vec{B}}{\partial t} \cdot d\vec{S}$	$\nabla \times \vec{E} = -\frac{\partial \vec{B}}{\partial t}$
II	$\oint_C \vec{H} \cdot d\vec{l} = \int_S \vec{J} \cdot d\vec{S}$	$\nabla \times \vec{H} = \vec{J}$
III	$\oint_\Sigma \vec{B} \cdot d\vec{S} = 0$	$\nabla \cdot \vec{B} = 0$
IV	$\oint_\Sigma \vec{D} \cdot d\vec{S} = \int_\tau \rho_v \, d\tau$	$\nabla \cdot \vec{D} = \rho_v$
Continuity equation	$\oint_\Sigma \vec{J} \cdot d\vec{S} = 0$	$\nabla \cdot \vec{J} = 0$

voltage source (or AC voltage). Thus, it is possible to demonstrate this assertion that electric currents are also sinusoidal time-dependent with the same frequency of feeding voltage, which is 50 Hz or 60 Hz depending on the country. In this case, considering the linear medium—where its physical properties are constant—the complex notation can be applied for representing the values involved in the phenomenon.

For such, we should remember that given a time-dependent sinusoidal value, we can express it by

$$y(t) = \sqrt{2}Y \cos[\omega t + \alpha], \qquad (3.52)$$

where

(Y) is its root mean square (or R.M.S.) value
$\omega = 2\pi f$ is the angular frequency
α is its phase

These kind of functions can be represented using the complex notation and applying Euler's identity, which establishes

$$e^{j\theta} = \cos\theta + j \, sen \, \theta$$

by the association of $y(t)$ to a complex function $\dot{y}(t)$ so that

$$\dot{y}(t) = \sqrt{2}Y \cos[\omega t + \alpha] + j\sqrt{2}Y \, sen[\omega t + \alpha].$$

Applying Euler's identity, it is possible to write

$$\dot{y}(t) = \sqrt{2}\dot{Y}e^{j\omega t},$$

where the term

$$\dot{Y} = Ye^{j\alpha} = Y\underline{/\alpha}$$

is called the phasor of $y(t)$.

Due to these considerations, we can write

$$y(t) = \text{Re}[\dot{y}(t)],$$

where Re[*] is a linear operator that selects the real part of [*].

Note that Re[*] is a linear operator because it presents the following properties:

1. $\text{Re}[\dot{Z}_1 + \dot{Z}_2] = \text{Re}[\dot{Z}_1] + \text{Re}[\dot{Z}_2]$
2. $\text{Re}[a\dot{Z}] = a\,\text{Re}[\dot{Z}]$, with a Real.

This procedure can be extended for vector field representations, where each components are sinusoidal time dependent.

For example, the vector field

$$\vec{v}(t) = \sqrt{2}V_x \cos[\omega t + \alpha_x]\vec{u}_x + \sqrt{2}V_y \cos[\omega t + \alpha_y]\vec{u}_y + \sqrt{2}V_z \cos[\omega t + \alpha_z]\vec{u}_z \qquad (3.53)$$

can be written as

$$\vec{v}(t) = \mathrm{Re}[\sqrt{2}\dot{\vec{V}}e^{j\omega t}],$$

where

$$\dot{\vec{V}} = \dot{V}_x\vec{u}_x + \dot{V}_y\vec{u}_y + \dot{V}_z\vec{u}_z,$$

with

$$\dot{V}_x = V_x e^{j\alpha_x}, \quad \dot{V}_y = V_y e^{j\alpha_y}, \quad \text{and} \quad \dot{V}_z = V_z e^{j\alpha_z}.$$

Here $\dot{\vec{V}}$ is named as complex vector.

The power of this mathematical tool is presented in differential equations, where quantities are sinusoidal time dependent.

Considering a differential equation involving a sinusoidal time-dependent quantity like

$$\frac{dy(t)}{dx} + a\frac{dy(t)}{dt} = g(t), \qquad (3.54)$$

where $y(t)$ is given by Equation 3.52 and

$$g(t) = \sqrt{2}G\cos[\omega t + \beta].$$

Considering our interest only in the *steady-state solution* of the previous differential equation, we can solve it applying the complex notation. The differential equation can be written in the following form:

$$\frac{d}{dx}\mathrm{Re}\left[\sqrt{2}\dot{Y}e^{j\omega t}\right] + a\frac{d}{dt}\mathrm{Re}\left[\sqrt{2}\dot{Y}e^{j\omega t}\right] = \mathrm{Re}\left[\sqrt{2}\dot{G}e^{j\omega t}\right]$$

or as

$$\mathrm{Re}\left[\sqrt{2}\frac{d\dot{Y}}{dx}e^{j\omega t}\right] + \mathrm{Re}\left[a\sqrt{2}\dot{Y}\frac{de^{j\omega t}}{dt}\right] = \mathrm{Re}\left[\sqrt{2}\dot{G}e^{j\omega t}\right].$$

Rearranging the terms on the left-hand side, we obtain

$$\mathrm{Re}\left[\sqrt{2}\frac{d\dot{Y}}{dx}e^{j\omega t} + j\omega a\sqrt{2}\dot{Y}e^{j\omega t}\right] = \mathrm{Re}\left[\sqrt{2}\dot{G}e^{j\omega t}\right].$$

When identifying the arguments, it results in

$$\frac{d\dot{Y}}{dx} + j\omega a \dot{Y} = \dot{G}. \tag{3.55}$$

If we compare the differential equation obtained from the original one (3.54), the following changes are observed:

1. The first term where the derivative appears related to the x position is applied to the phasor $y(t)$,
2. The second term where the time derivative (d/dt) of Equation 3.54 is replaced by the product $j\omega$ and the unknown by its phasor.
3. The the right-side term where the function is replaced by its respective phasor.

If the differential equation has any term involving the time integral of the unknown, it is possible to demonstrate that this would be replaced on the equation transformed by the complex constant $1/j\omega$.

The bigger advantage is that in the differentiated transformed equation both time-dependent derivatives and integrals are replaced by non-time-dependent terms. In case of differential equations involving only time-dependent derivatives or integrals, the transformed equation will be a simple algebraic equation.

3.8.4 MAGNETODYNAMICS

We characterize magnetodynamics state (or quasi-static state) when all quantities are sinusoidal time dependent with a frequency small enough to neglect displacement currents, that is, $\partial \vec{D}/\partial t = 0$. Table 3.5 describes Maxwell's equations suitable for solving any magnetodynamic problem using complex notation.

TABLE 3.5
Maxwell's Equations on Magnetodynamics

Equation	Integral Form	Differential Form
I	$\oint_C \dot{\vec{E}} \cdot d\vec{l} = -j\omega \int_S \dot{\vec{B}} \cdot d\vec{S}$	$\nabla \times \dot{\vec{E}} = -j\omega \dot{\vec{B}}$
II	$\oint_C \dot{\vec{H}} \cdot d\vec{l} = \int_S \dot{\vec{J}} \cdot d\vec{S}$	$\nabla \times \dot{\vec{H}} = \dot{\vec{J}}$
III	$\oint_\Sigma \dot{\vec{B}} \cdot d\vec{S} = 0$	$\nabla \cdot \dot{\vec{B}} = 0$
IV	$\oint_\Sigma \dot{\vec{D}} \cdot d\vec{S} = \int_\tau \dot{\rho}_v \, d\tau$	$\nabla \cdot \dot{\vec{D}} = \dot{\rho}_v$
Continuity equation	$\oint_\Sigma \dot{\vec{J}} \cdot d\vec{S} = 0$	$\nabla \cdot \dot{\vec{J}} = 0$

TABLE 3.6
Maxwell's Equations in Free Space

Equation	Integral Form	Differential Form
I	$\oint_c \vec{E} \cdot d\vec{l} = -\int_s \dfrac{\partial \vec{B}}{\partial t} \cdot d\vec{S}$	$\nabla \times \vec{E} = -\dfrac{\partial \vec{B}}{\partial t}$
II	$\oint_c \vec{H} \cdot d\vec{l} = \int_s \dfrac{\partial \vec{D}}{\partial t} \cdot d\vec{S}$	$\nabla \times \vec{H} = \dfrac{\partial \vec{D}}{\partial t}$
III	$\oint_\Sigma \vec{B} \cdot d\vec{S} = 0$	$\nabla \cdot \vec{B} = 0$
IV	$\oint_\Sigma \vec{D} \cdot d\vec{S} = 0$	$\nabla \cdot \vec{D} = 0$

3.8.5 Fast Time-Dependent Electromagnetic Fields

The most important contribution of Maxwell toward electromagnetism was predicting the existence of electromagnetic waves in fast time-dependent electromagnetic fields not only in free space but also in a material media. Indeed in case of ideal dielectrics free of electric charge, that is, $J = \rho_v = 0$, as the free space, we can rewrite Maxwell's equations, as shown in Table 3.6.

From this set of equations, it is possible to define the property of propagation of electromagnetic fields, a characteristic of electromagnetic waves [47].

3.9 BOUNDARY CONDITIONS

The presence of electric or magnetic fields is quite common in electrical devices that are produced in one medium and transmitted to another with completely different physical properties. This passage from one medium to another implies amplitude and directional changes of fields due to the difference of its physical properties. Our objective is to evaluate variations felt by electromagnetic fields (electric and magnetic) under this change of medium. For such, it is convenient to divide fields of both medium into its normal and tangential components on the interface between any two dissimilar medium, as shown in Figure 3.20.

Thus, a given electromagnetic field $\vec{V}_1$ produced in medium 1, with physical properties ε_1, μ_1, and σ_1, emerges in medium 2, with physical properties ε_2, μ_2, and σ_2, assuming a new value $\vec{V}_2$.

Dividing these vectors into its both normal and tangential components to the surface, we obtain the following:

$$\vec{V}_1 = \vec{V}_{n1} + \vec{V}_{t1} \tag{3.56}$$

$$\vec{V}_2 = \vec{V}_{n2} + \vec{V}_{t2}. \tag{3.57}$$

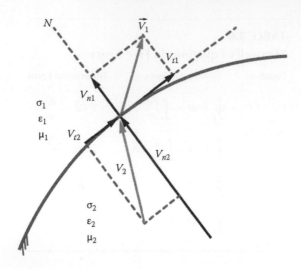

FIGURE 3.20　Normal and tangential components of the vector.

3.9.1　Normal Components of $\vec{B}$ and $\vec{H}$

Figure 3.21 shows the interface of two material medium, where a flux density field $\vec{B}_1$ produced in medium 1 emerges in medium 2 assuming the value $\vec{B}_2$.

Consider an interface point involved by a very small box located at its center, as shown in Figure 3.21. As dimensions Δx, Δy, and Δz are infinitesimal, we can consider that magnetic fields are constant on each side of the surface. Let us now apply

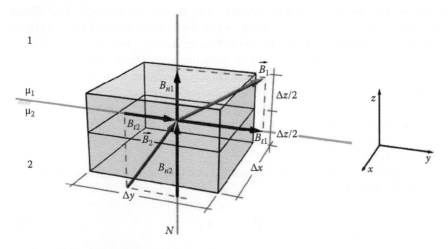

FIGURE 3.21　Boundary conditions for $\vec{B}$.

Maxwell's third equation to this closed surface, remembering that the elementary area vector $d\vec{S}$ is normal to the surface. So, we have

$$\oint_{\Sigma} \vec{B} \cdot d\vec{S} = 0. \tag{3.58}$$

As $\vec{B}$ is constant inside of Σ, Maxwell's third equation is reduced to

$$\sum_{i=1}^{6} \vec{B}_i \cdot \Delta\vec{S}_i = 0, \tag{3.59}$$

where

i represents the sides of the surface
$\Delta\vec{S}_i$ is the elementary surface vector of the corresponding side

Note that only the normal components of the flux density fields contribute to the magnetic flux on the closed surface through the upper and lower lids because the scalar product $\vec{B}_i \cdot \Delta\vec{S}_i$ is null on its other sides.

Thus, we obtain

$$-B_{n2}\Delta x\Delta y + B_{n1}\Delta x\Delta y = 0 \tag{3.60}$$

or

$$B_{n1} = B_{n2}. \tag{3.61}$$

The result expressed in Equation 3.61 evidences the fact that the normal component of the flux density field is the continuous on the interface of the material media.

Remembering the constitutive relation $\vec{B} = \mu\vec{H}$, the relation of the normal components of the magnetic field vector is written as follows:

$$\mu_1 H_{n1} = \mu_2 H_{n2}. \tag{3.62}$$

This evidences the discontinuity of the normal component of the magnetic field in an interface of two different medium.

3.9.2 Normal Components of $\vec{D}$ and $\vec{E}$

The geometry for the analysis of behavior of the normal component of the displacement vector and electric field is the same as shown in Figure 3.21, and it is enough to replace $\vec{B}$ by $\vec{D}$. However, Maxwell's fourth equation in an integral form is applied for obtaining the behavior of the normal components of the displacement vector $\vec{D}$.

$$\oint_{\Sigma} \vec{D} \cdot d\vec{S} = \int_{\tau} \rho_v \, d\tau. \tag{3.63}$$

Related to the left-hand side the result is identical to that of Equation 3.60, that is,

$$\oint_{\Sigma} \vec{D} \cdot d\vec{S} = -D_{n2}\Delta x\Delta y + D_{n1}\Delta x\Delta y. \qquad (3.64)$$

The right-hand side of Equation 3.63 corresponds to the total amount of electric charges inside the closed surface. As the box dimensions are elementary, the quantity of electric charges inside the closed surface consists of an eventual quantity of electric charges deposited on the interface of both media, that is,

$$\int_{\tau} \rho_v \, d\tau = \Delta q_S. \qquad (3.65)$$

Comparing the results obtained in Equations 3.64 and 3.65, we obtain

$$-D_{n2}\Delta x\Delta y + D_{n1}\Delta x\Delta y = \Delta q_S \qquad (3.66)$$

or

$$D_{n1} - D_{n2} = \frac{\Delta q_S}{\Delta x\Delta y}. \qquad (3.67)$$

At a limit of Δx, $\Delta y \rightarrow 0$, the right-hand side of Equation 3.67 indicates the superficial density of electric charges on the interface of both media; thus, it can be written as following:

$$D_{n1} - D_{n2} = \rho_S. \qquad (3.68)$$

This result evidences the discontinuity of the normal component of the displacement vector on the interface of both media with electric charge distribution on the surface of this interface.

From the constitutive relation $\vec{D} = \varepsilon \vec{E}$ the behavior of the normal component of the electric field on the interface of both media is obtained, that is,

$$\varepsilon_1 E_{n1} - \varepsilon_2 E_{n2} = \rho_S. \qquad (3.69)$$

3.9.3 Normal Components of $\vec{J}$

The behavior of the normal component of the current density vector is derived from Equation 3.24, which is reproduced here by convenience as

$$\oint_{\Sigma} \left(\vec{J} + \frac{\partial \vec{D}}{\partial t} \right) \cdot d\vec{S} = 0$$

or as

$$\oint_{\Sigma} \vec{J} \cdot d\vec{S} = -\frac{d}{dt} \oint_{\Sigma} \vec{D} \cdot d\vec{S}. \tag{3.70}$$

Using the same geometry of Figure 3.21, we can write

$$-J_{n2}\Delta x\Delta y + J_{n1}\Delta x\Delta y = -\frac{d}{dt}\left(-D_{n1}\Delta x\Delta y + D_{n2}\Delta x\Delta y\right).$$

Although they are small, Δx and Δy are non-null resulting in

$$-J_{n2} + J_{n1} = -\frac{d}{dt}\left(-D_{n1} + D_{n2}\right).$$

Remembering that

$$D_{n1} - D_{n2} = \rho_S,$$

we finally obtain

$$J_{n1} - J_{n2} = -\frac{d\rho_S}{dt}. \tag{3.71}$$

This result shows that the normal component of the current density vector $\vec{J}$ is discontinuous of a value equal to the time-variation rate of the superficial density of charges deposited on the interface of both media.

3.9.4 TANGENTIAL COMPONENTS OF $\vec{H}$

Figure 3.22 shows a magnetic vector $\vec{H}_1$ produced in medium 1 that emerges in medium 2 assuming a new value $\vec{H}_2$.

For establishing the behavior of the tangential component of the magnetic field $\vec{H}$, we will work with Maxwell's second equation given by

$$\oint_C \vec{H} \cdot d\vec{l} = \int_S \left(\vec{J} + \frac{\partial \vec{D}}{\partial t}\right) \cdot d\vec{S}. \tag{3.72}$$

For such, an elementary rectangular closed path with dimensions $\vec{H}_1$ and $\vec{H}_2$ can be considered constant along its edges. The left-hand side of Equation 3.72 applied to the elementary rectangular closed path can be written as following:

$$\oint_C \vec{H} \cdot d\vec{l} = \sum_{i=1}^{4} \vec{H}_i \cdot \Delta \vec{l}_i, \tag{3.73}$$

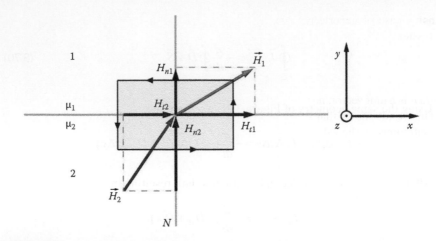

FIGURE 3.22 Boundary conditions for $\vec{H}$.

where

i corresponds to the side of the infinitesimal rectangular closed path

$\vec{H}_i$ represents the components (normal and tangential) of the vector H on this same side

Note that only the tangential components contribute to $\vec{H}$ *circulation* on the referred closed path because other scalar products result null so that

$$\oint_C \vec{H} \cdot d\vec{l} = H_{t1}\Delta x - H_{t2}\Delta x. \tag{3.74}$$

The right-hand side corresponds to the linkage current by the closed path. As our closed path is infinitesimal, that is, $\Delta y \to 0$, this leakage current corresponds to an eventual current existing on the interface of both media, thus

$$\int_S \left(\vec{J} + \frac{\partial \vec{D}}{\partial t} \right) \cdot d\vec{S} = \Delta I_S. \tag{3.75}$$

Comparing expressions (3.74) and (3.75), we obtain

$$H_{t1}\Delta x - H_{t2}\Delta x = \Delta I_S \tag{3.76}$$

or

$$H_{t1} - H_{t2} = \frac{\Delta I_S}{\Delta x}. \tag{3.77}$$

The right-hand side of Equation 3.77, where $\Delta x \to 0$, is the z component of the superficial current density vector $\vec{J}_l$.

For a case of generic positioning of fields related to the reference system, the expression (3.77) can be written as following:

$$\vec{n} \times \left(\vec{H}_2 - \vec{H}_1 \right) = \vec{J}_l, \tag{3.78}$$

where $\vec{n}$ is a unit vector, normal to the surface, directed from medium 1 to medium 2.

This result shows that the tangential component of the magnetic field vector is discontinuous on the value corresponding to the superficial density of current on the interface of both media.

From the constitutive relation $\vec{H} = \vec{B} / \mu$, it is possible to obtain the behavior of the tangential components of the magnetic field resulting in

$$\vec{n} \times \left(\frac{\vec{B}_2}{\mu_1} - \frac{\vec{B}_1}{\mu_2} \right) = \vec{J}_l. \tag{3.79}$$

3.9.5 Tangential Components of $\vec{E}$

The determination of the behavior of tangential components of an electric field is obtained by applying Maxwell's first equation to a rectangular closed path in a similar method as shown in Figure 3.22.

Thus, similar to the previous example, we can write

$$\oint_C \vec{E} \cdot d\vec{l} = E_{t1}\Delta x - E_{t2}\Delta x. \tag{3.80}$$

The right-hand side of Maxwell's first equation, which corresponds to EMF induced in the closed path, is null. Indeed, the area of the closed path tends to be zero for $\Delta x, \Delta y \to 0$, that is,

$$-\int_S \frac{\partial B}{\partial t} \to 0. \tag{3.81}$$

Thus, we can write

$$E_{t1} = E_{t2}. \tag{3.82}$$

However, the tangential component of the electric field is continuous on the separation interface of both media.

3.9.6 Tangential Components of $\vec{J}$

The behavior of the tangential component of the current density vector is obtained from Equation 3.82 from the constitutive relation $\vec{E} = \vec{J}/\sigma$; thus, we can write

$$\frac{J_{t1}}{\sigma_1} = \frac{J_{t2}}{\sigma_2}. \tag{3.83}$$

TABLE 3.7

Boundary Conditions

Vector	Normal Components	Tangential Components
$\vec{B}$	$B_{n1} = B_{n2}$	$\vec{n} \times \left(\dfrac{\vec{B}_2}{\mu_1} - \dfrac{\vec{B}_1}{\mu_2} \right) = \vec{J}_t$
$\vec{H}$	$\mu_1 H_{n1} = \mu_2 H_{n2}$	$\vec{n} \times \left(\vec{H}_2 - \vec{H}_1 \right) = \vec{J}_t$
$\vec{D}$	$D_{n1} - D_{n2} = \rho_S$	$\dfrac{D_{t1}}{\varepsilon_1} = \dfrac{D_{t2}}{\varepsilon_2}$
$\vec{E}$	$\varepsilon_1 E_{n1} - \varepsilon_2 E_{n2} = \rho_S$	$E_{t1} = E_{t2}$
$\vec{J}$	$J_{n1} - J_{n2} = -\dfrac{d\rho_S}{dt}$	$\dfrac{J_{t1}}{\sigma_1} = \dfrac{J_{t2}}{\sigma_2}$

Note: $\vec{n}$ is a unit vector, which is normal to surface and directed from medium 1 to medium 2.

On the other hand, the behavior of the tangential component of the displacement vector is also obtained from Equation 3.82 from the constitutive relation $\vec{E} = \vec{D}/\varepsilon$. It results in

$$\frac{D_{t1}}{\varepsilon_1} = \frac{D_{t2}}{\varepsilon_2}. \tag{3.84}$$

In conclusion, Table 3.7 shows all boundary conditions obtained from the application of Maxwell equations in an infinitesimal domains that enable us to evaluate the effect of medium changing in electromagnetic fields.

3.10   ELECTROMAGNETISM POTENTIALS

Although Maxwell's equations give all information necessary for the complete identification of an electromagnetic phenomenon, they present situations when solutions are extremely complex. For an easy way of obtaining a solution, scalar functions (or vector fields), said potential, are defined that simplify its search.

The concepts of potentials (scalar or vector) are largely explored in the mathematical formalism of numerical methods. In electromagnetism, both electrical and magnetic potentials are powerful tools for solving numerically both Laplace and Poisson equations.

Although they have no physical meaning, electromagnetism potentials, due to their constant utilization, are confounded in some cases to the electric or magnetic fields that cause them. It is common to attribute our pain due to an electric shock to the voltage of 110 V from the power outlet, when all the damage was caused by the electric field to which we were subjected to when we neglect while dealing with electricity.

3.10.1   THE SCALAR ELECTRIC POTENTIAL—V

The scalar electric potential is established for the electric field in electrostatic studies, for which Maxwell's first equation in differential form is reduced to the following:

$$\nabla \times \vec{E} = 0. \tag{3.85}$$

From the properties of the field operators, we obtain

$$\nabla \times \nabla f = 0, \tag{3.86}$$

which establishes that the curl of the gradient of any scalar function is identically null.

Thus, the electric field of Equation 3.85 is a derivate of the gradient of a function called as electric potential function so that

$$\vec{E} = -\nabla V. \tag{3.87}$$

The negative sign was placed by convenience. Electric potential functions so defined is such that the difference between their values obtained at two different points is named voltage between these points, which can be measured by a voltmeter under certain conditions.

The potential definition through this procedure does not assure uniqueness to V because the function

$$V' = V + \text{Constant}$$

also takes the same electric field. To deal with this difficulty, it is assured that the uniqueness of V arbitrating a value for the electric potential function at a point in space, that is, is selected as a reference to potentials. It is common, always when possible, to attribute null electric potential to a remote point. A potential function with this characteristic is only possible to be selected in case of conservative fields, that is, the ones having null curl.

3.10.2   THE SCALAR MAGNETIC POTENTIAL—Ψ

When the magnetic field occurs in a region free of electric current ($J = 0$), the scalar magnetic potential function can also be defined, as we did with the electric potential function. Under this condition, Maxwell's second equation is reduced to

$$\nabla \times \vec{H} = 0. \tag{3.88}$$

Thus, the scalar magnetic potential function is defined as the function ψ so that

$$\vec{H} = -\nabla \psi. \tag{3.89}$$

It is important to observe that the magnetic scalar potential function is not applied to regions where there is electric current because in this case J is not null.

3.10.3 The Magnetic Potential Vector—$\vec{A}$

The magnetic potential vector is associated to the magnetic flux density in the presence of electric current due to the inherent property described in Maxwell's third equation to which

$$\nabla \cdot \vec{B} = 0. \tag{3.90}$$

From the identities of the field operators, we obtain

$$\nabla \cdot \nabla \times \vec{V} = 0, \tag{3.91}$$

which establishes that the divergence of the curl of any vector field is identically null.

Thus, the magnetic field of Equation 3.90 is a derivate of the curl of a vector field called magnetic potential vector represented as follows:

$$\vec{B} = \nabla \times \vec{A}. \tag{3.92}$$

The definition of the magnetic potential vector from this procedure does not assure the uniqueness to $\vec{A}$ because the vector field

$$\vec{A}' = \vec{A} + \nabla f,$$

with f being any scalar function, which also takes the same magnetic field.

For assuring this uniqueness, its divergence is also imposed selecting

$$\nabla \cdot \vec{A} = 0. \tag{3.93}$$

A vector field with these characteristics is only possible to be selected in case of solenoid fields, that is, the ones having null divergence or closed field lines.

3.10.4 The Electric Potential Vector

The electric potential vector is associated to the current density vector in stationary regime.

Applying the same development in the definition of the magnetic potential vector through the third Maxwell equation, we can perform with the continuity equation

$$\nabla \cdot \vec{J} = 0. \tag{3.94}$$

A procedure similar to the one used for the magnetic flux density field can be used here to indicate that $\vec{J}$ is a derivate of the curl of a vector called electric potential vector ($\vec{T}$) so that

$$\vec{J} = \nabla \times \vec{T}. \tag{3.95}$$

For assuring this uniqueness, its divergence is also imposed selecting

$$\nabla \cdot \vec{T} = 0. \tag{3.96}$$

3.10.5 POTENTIALS IN TIME-DEPENDENT FIELDS

For a time-dependent electric field, it is not possible to define the scalar electric potential as shown in Equation 3.87 without some mathematical manipulations because the time-dependent electric field is a nonconservative vector field.

From Maxwell's first equation in a differential form and definition of the vector magnetic potential, which we reproduce here by convenience

$$\nabla \times \vec{E} = -\frac{\partial \vec{B}}{\partial t}$$

$$\vec{B} = \nabla \times \vec{A},$$

we can write

$$\nabla \times \vec{E} = -\frac{\partial \nabla \times \vec{A}}{\partial t}$$

or as

$$\nabla \times \left(\vec{E} + \frac{\partial \vec{A}}{\partial t} \right) = 0.$$

Since there is a null curl, it can be defined, for example, as the procedure for obtaining Equation 3.87—the scalar potential function V as

$$\vec{E} + \frac{\partial \vec{A}}{\partial t} = -\nabla V$$

or as

$$\vec{E} = -\nabla V - \frac{\partial \vec{A}}{\partial t}. \tag{3.97}$$

We can conclude the result obtained, that is, the time-dependent electric field has two terms: one due to the separation of electric charges represented as $-\nabla V$ and known as impressed electric field with null curl, and the other one due to the time-dependent electric current represented as $-\partial \vec{A} / \partial t$ and known as induced electric field with non-null curl. This result evidences the nonuniqueness of the voltage between two points in case of time-dependent electric field, because the line integration of the induced electric field associated to the second term of (3.97) depends on the path adopted for the integration of the electric field to obtain electric voltage.

4 Finite Element Method in Static, Electric, and Magnetic Fields

4.1 INTRODUCTION

As mentioned in the Chapter 1, the finite element method (FEM) is presented in most publication, as the one of Zienkiewics [2], through complex mathematical formulations, frequently inaccessible to or difficult to understand for both undergraduate students and beginners. The discussions about mathematics involved FEM are constrained to the academic environment.

Aiming to formulate an easy methodology for undergraduate students from the Polytechnic School of the University of São Paulo and encouraging them to be involved with FEM, we developed in the 1980s a methodology based exclusively on the integration of Maxwell's equations, which avoids complex and difficult mathematical treatments [11,36].

This formulation is also a powerful tool to introducing this numeric method to professionals involved in electric equipment designs. FEM is recognized as an important instrument for improving the quality of the products, enabling the professional explores more alternatives in short time.

Two classical mathematical formulations of FEM are presented in current textbooks. The first one is based on the minimization of a functional energy. In this case, we have to search for a solution that minimizes an electric energy integral function involved in the phenomenon related to state variables, which are unknown in the problem of nodes from finite elements mesh. The second one is based on the weighted residual method, where there is a need for an approximated solution of Poisson's equation, subject to well-established boundary conditions through numeric techniques, which are very complex and and without an association with any physical phenomenon making it hard to understand. A well-known technique is the Galerkin technique, which, in most problems, generates a system of symmetric linear equations, making it convenient.

Although they are very different, these two mathematical formulations give the same result in two-dimensional and static problems. In problems involving time-dependent phenomena, the functional energy minimization method is rarely applied, and there is no alternative to the weighted residual methods [27,32].

Figure 4.1 shows the possible ways for arriving at the final equations system of FEM from classical formulations using our method. Note that, unlike other methods, this one arrives at the same results of the ones obtained by both the Garlerkin technique and variation formulation without the manipulation of Maxwell's equations through a very simple mathematical treatment, as we will see.

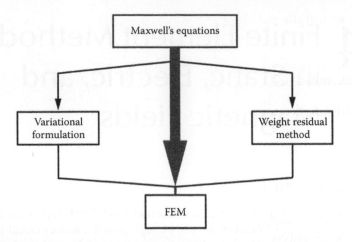

FIGURE 4.1 The possible ways for FEM analysis in electromagnetics.

4.2 STEPS IN FEM APPLICATION

The first step in FEM application is selecting the type of formulation: two dimensional or three dimensional. Let us start our discussion by understanding first the two-dimensional finite element method (2D-FEM) because it is easier and has been known to solve 90% of the important problems in electric engineering.

Once the domain of study is identified, we use the triangular finite element with straight sides, called first-order element, to discretize this domain. The choice of the triangular element is because the mathematical formulation of FEM results very simple bonded to the fact that it easy the implantation of automatic generation routines of triangular finite elements meshes.

With regard to the selection of the interpolation function of the state variable on the interior of the element, let us restrict, for this case, ourselves to the linear interpolation function, because it helps with a simple integration of Maxwell's equations.

Now, we will start with the static regime phenomenon, where $\partial/\partial t = 0$. Thus, we will adhere to the following sequence of plans and axisymmetric problems:

- Electrostatics
- Electrokinetics
- Magnetostatics

For the quasi-static time-dependent phenomena, where the displacement current is neglected but not the conduction current, we will discuss the following:

- Magnetodynamics in sinusoidal time-dependent steady state
- Magnetodynamics in transient regime

We will also look into high-frequency electromagnetic phenomena, also called fast, time-dependent electromagnetic fields regime. In this case, the mathematical formulation of 2D-FEM will be presented for wave propagation studies on *plane wave guides*.

Finally, we will also make a short incursion into the mathematical formulation of the 3D-FEM in studies on

- 3D Electrostatics
- 3D Electrokinetics

We will discuss how to impose the following constraints in boundary conditions:

- Dirichlet conditions
- Neumann conditions
- Periodicity conditions
- Floating conditions

These boundary conditions cover almost all constraints to the fields in electric devices. Other special conditions that can reduce the size of a domain or stabilize the numerical processes are beyond the scope of this text.

The next step consists in the assembly of the equations system, the solution of which will supply the potential distribution—electric, magnetic, or both—from which we will have the conditions for obtaining the electric or magnetic field distribution and the quantities associated with the phenomenon.

The FEM principle consists in subdividing the domain under study, which we will suppose as being two dimensional within smaller subdomains called finite elements. This subdivision can be made using different types of polygons; however, as it was already mentioned, let us assume the domain is subdivided into triangular finite elements.

Figure 4.2 shows a cross section of a two-dimensional domain of unitary depth divided into triangular elements, which are numbered from 1 to NE. The vertexes of these elements, called nodes from the finite element mesh, are numbered from 1 to NN.

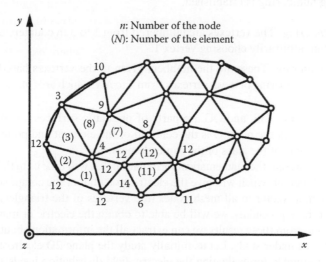

FIGURE 4.2 Two-dimensional domain divided into triangular elements.

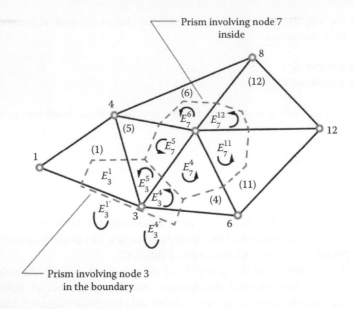

FIGURE 4.3 Prisms involving nodes.

Let us take each domain's node by a polygonal cross-sectional prism constituted by the union of straight segments that connect the centroids to the middle of the side of the elements. We will also suppose that this prism's depth is unitary.

Figure 4.3 shows a finite element mesh detail highlighting two prisms, one involving an internal node to the mesh (node 7) and the other involving a boundary node from the domain (node 3).

Figure 4.4 highlights a generic finite element extracted from the domain to which the following numbering is established.

Local numbering: The vertexes are numbered from 1 to 3 in counterclockwise direction, arbitrarily choosing vertex 1.

Global numbering: These are the numbers given to the vertexes based on the arbitrary numbering of mesh vertexes from 1 to NN, which are (p), (q), and (r).

The segments $\overline{PO}$, $\overline{OS}$, and $\overline{OG}$ are parts of prismatic surfaces involving nodes 1, 2, and 3. Note that the point O is the centroid of the element and points P, S, and G the middle points of the sides of the same element.

We will now present the mathematical formulation that will take us to the equations system, the solution of which will give the electric potential or the component z of the magnetic potential vector in all mesh nodes (the vertexes of the triangles). From the knowledge of these potentials, we will be able to obtain the electric or magnetic field distribution, and from these results we can extract all the information about the operation of the device under study. Let us initially study the plane 2D electrostatics. This formulation is suitable for evaluating the electric field distribution inside the devices that its cross section is repeated along its depth in parallel planes to this section.

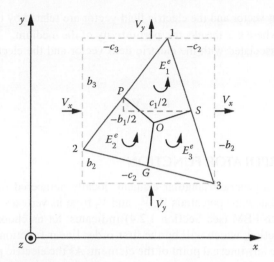

FIGURE 4.4 Generic finite element.

4.3 2D ELECTROSTATICS

In electrostatics, our objective is to establish both the electric potential and electric field distribution produced by a set of conductors, excited by voltage sources, and each separated from the other by ideal dielectric medium (homogeneous or not) with or without electric charges (Figure 4.5).

The starting point for this development is Maxwell's fourth equation (Gauss' law of electrostatics):

$$\oint_{\Sigma} \vec{D} \cdot d\vec{S} = Q_i \tag{4.1}$$

where

$\vec{D}$ is the displacement vector (C/m²)

Q_i is the total amount of internal electrical charge to surface Σ

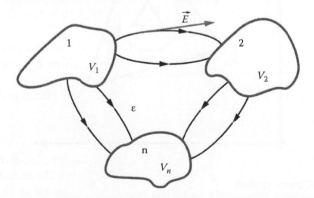

FIGURE 4.5 Typical problem of electrostatics—Excited conductors separated by dielectric.

Both displacement vector and the electric field vector are related by the constitutive relation $\vec{D} = \varepsilon\vec{E}$, where ε is the electric permittivity of the medium.

The relation associated with the electric field vector and the electrical potential function is

$$\vec{E} = -\nabla V \tag{4.2}$$

4.4   2D INTERPOLATOR FUNCTION

Figure 4.6 shows a generic triangular element, locally numbered from 1 to 3, for which we should know the potentials V_1, V_2, and V_3 from its vertexes.

As Step 4 from FEM (see Section 1.2.4) indicates, let us choose the simpler interpolation function, which will be the first-order linear function. Here, $R(x, y)$ is used to indicate any internal point of the element. As the electric potential function is a continuous function, we can express the electric potential on that point using the following function:

$$V = \alpha_1 + \alpha_2 x + \alpha_3 y \tag{4.3}$$

where coefficients α_1, α_2, and α_3 are related to V_1, V_2, and V_3.

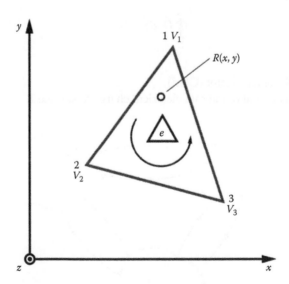

FIGURE 4.6   Generic element.

To establish these coefficients, it is enough to replace (x) and (y) from (4.3) with the coordinates of the element's vertexes (or nodes), which results in the following equations system:

$$V_1 = \alpha_1 + \alpha_2 x_1 + \alpha_3 y_1$$
$$V_2 = \alpha_1 + \alpha_2 x_2 + \alpha_3 y_2 \quad (4.4)$$
$$V_3 = \alpha_1 + \alpha_2 x_3 + \alpha_3 y_3$$

The solution of (4.4) related to the coefficients α's results in

$$\alpha_1 = \frac{1}{2\Delta}\left(a_1 V_1 + a_2 V_2 + a_3 V_3\right)$$

$$\alpha_2 = \frac{1}{2\Delta}\left(b_1 V_1 + b_2 V_2 + b_3 V_3\right) \quad (4.5)$$

$$\alpha_3 = \frac{1}{2\Delta}\left(c_1 V_1 + c_2 V_2 + c_3 V_3\right)$$

where

$$a_1 = x_2 y_3 - x_3 y_2; \quad b_1 = y_2 - y_3; \quad c_1 = x_3 - x_2 \quad \text{and} \quad \Delta = (b_1 c_2 - b_2 c_1)/2 \quad (^*)$$

Replacing (4.5) in (4.3), we obtain the electric potential's expression at any internal point of the element, through a linear interpolation of the potentials on their nodes, as follows:

$$V(x,y) = N_1 V_1 + N_2 V_2 + N_3 V_3 \quad (4.6)$$

where

$$N_i = \frac{1}{2\Delta}\left(a_i + b_i x + c_i y\right) \quad i = 1, 2, 3$$

The function $N_i(x, y)$ is the element's shape function, which is expressed as follows:

$$N_i(x_j, y_j) = \delta_{ij} \quad (4.7)$$

where δ_{ij} is Kronecker's symbol and is expressed as

$$\delta_{ij} = \begin{cases} 1 & \text{if } i = j \\ 0 & \text{if } i \neq j \end{cases}$$

---

* The others coefficients a, b, and c are obtained by cyclic rotation of their indexes and Δ is the area of the element.

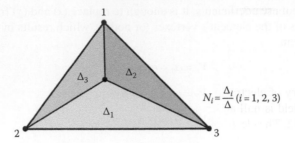

FIGURE 4.7   Geometric interpretation of the interpolation function.

Figure 4.7 presents a geometric interpretation of this function.

It can be observed that using this type of interpolation, any curved surface is approximated by a tiled surface with triangular elements; hence, the larger the number of elements, in principle, the better the accuracy in the representation of the referred surface.

4.5   ELEMENT'S MATRIX IN ELECTROSTATICS

Using $\vec{E} = -\nabla V$, each component of the electric field vector can be expresses as follows:

$$E_x = -\frac{\partial V}{\partial x} = -\frac{1}{2\Delta}\left(b_1V_1 + b_2V_2 + b_3V_3\right)$$

$$E_y = -\frac{\partial V}{\partial y} = -\frac{1}{2\Delta}\left(c_1V_1 + c_2V_2 + c_3V_3\right)$$

$$(4.8)$$

An analysis of this expression shows that a linear interpolation of the electric potential function $V(x, y)$ results in a constant electric field inside the element—as expected!

The application of Maxwell's fourth equation to prismatic closed surfaces of unitary depth and cross section identical to the polygons shown in Figure 4.1 can be made by sections using the following procedure:

For the prism involving node 7, we can write

$$\oint_{\Sigma_7} \vec{D} \cdot d\vec{S} = E_7^4 + E_7^{11} + E_7^{12} + E_7^6 + E_7^5 \qquad (4.9)$$

where

$$E_i^e = \int_{S_i} \vec{D} \cdot d\vec{S} \qquad (4.10)$$

It represents the displacement vector flux at the sections of the surface Σ_i, which involves node (i) belonging to element (e).

In the case of node 3, which belongs to the boundary of the domain, we have

$$\oint_{\Sigma_3} \vec{D} \cdot d\vec{S} = E_3^4 + E_3^5 + E_3^1 + E_3^{1'} + E_3^{4'}$$

(4.11)

At boundary of electrostatics problems, we observe two different situations: (1) The electric field is null at all boundary conditions, so the last two terms of (4.11) are also null. (2) The electric field is null at a portion of the boundary and tangent in the remainder, so in the first portion the last two terms of (4.11) are null for reasons already presented and are null in the remaining portion because the electric field vector is normal to the elementary surface vector. Thus, for each finite element, three sections of the surface integrals can be calculated: an integral section of the surface that involves node 1 (E_1^e), one involving node 2 (E_2^e), and another involving node 3 (E_3^e), as shown in Figure 4.4.

As the electric field is constant inside the element, the calculation of (E_1^e) over that generic element is easy, simplifying the evaluation of the closed surface integral:

$$E_1^e = \int_{S_1} \vec{D} \cdot d\vec{S}$$

In this case, the indicated section surface S_1 is the one whose sides on the x, y plane are the segments $\overline{PO}$ and $\overline{OS}$ with unit depth on the z-axis.

As $\vec{D}$ and $d\vec{S}$ are given by

$$\vec{D} = \varepsilon E_x \vec{u}_x + \varepsilon E_y \vec{u}_y$$

$$d\vec{S} = -\Delta y \vec{u}_x - \Delta x \vec{u}_y$$

we get

$$E_1^e = \int_{S_1} \vec{D} \cdot d\vec{S} = -\varepsilon E_x \Delta y - \varepsilon E_y \Delta x$$

(4.12)

Replacing E_x and E_y by their values expressed in (4.8) and noting that

$$\Delta x = x_s - x_p = \frac{1}{2}(x_3 - x_2) = \frac{c_1}{2}$$

$$\Delta y = y_p - y_s = \frac{1}{2}(y_2 - y_3) = \frac{b_1}{2}$$

we get

$$E_1^e = \frac{\varepsilon}{4\Delta}\left[\left(b_1 b_1 + c_1 c_1\right)V_1 + \left(b_1 b_2 + c_1 c_2\right)V_2 + \left(b_1 b_3 + c_1 c_3\right)V_3\right]$$

(4.13)

The evaluation of E_2^e, which represents the displacement vector flux on the section of surface Σ_2 that involves node (2) belonging to the element (e) is obtained similarly to E_1^e, as follows:

$$E_2^e = \int_{S_2} \vec{D} \cdot d\vec{S} = \varepsilon E_x \Delta y + \varepsilon E_y \Delta x \qquad (4.14)$$

where

$$\Delta x = x_g - x_p = \frac{1}{2}(x_3 - x_1) = -\frac{c_2}{2}$$

$$\Delta y = y_p - y_g = \frac{1}{2}(y_1 - y_3) = -\frac{b_2}{2}$$

results in

$$E_2^e = \frac{\varepsilon}{4\Delta}\left[(b_2 b_1 + c_2 c_1)V_1 + (b_2 b_2 + c_2 c_2)V_2 + (b_2 b_3 + c_2 c_3)V_3\right] \qquad (4.15)$$

Following the analog procedure, the reader can easily deduce that

$$E_3^e = \frac{\varepsilon}{4\Delta}\left[(b_3 b_1 + c_3 c_1)V_1 + (b_3 b_2 + c_3 c_2)V_2 + (b_3 b_3 + c_3 c_3)V_3\right] \qquad (4.16)$$

In summary, the contributions of the displacement vector flux through surface sections involving nodes 1, 2, and 3 from element (e) can be expressed in matrix form as follows:

$$\begin{bmatrix} E_1^e \\ E_2^e \\ E_3^e \end{bmatrix} = \frac{\varepsilon}{4\Delta}\begin{pmatrix} b_1 b_1 + c_1 c_1 & b_1 b_2 + c_1 c_2 & b_1 b_3 + c_1 c_3 \\ b_2 b_1 + c_2 c_1 & b_2 b_2 + c_2 c_2 & b_2 b_3 + c_2 c_3 \\ b_3 b_1 + c_3 c_1 & b_3 b_2 + c_3 c_2 & b_3 b_3 + c_3 c_3 \end{pmatrix}\begin{bmatrix} V_1 \\ V_2 \\ V_3 \end{bmatrix} \qquad (4.17)$$

The squared matrix of this equation is called element's matrix and presents the symmetry and singularity characteristics.

The second member of Maxwell's fourth equation is equal to the internal charge to the surface. Thus, for the closed surface (prism) involving node 7 from Figure 4.2, we can write

$$Q_7 = Q_7^4 + Q_7^{11} + Q_7^{12} + Q_7^6 + Q_7^5$$

where Q_i^e is the portion of the total amount of charge contained in the interior volume with cross section Δ and unitary height that involves node (i) belonging to element (e).

Taking the generic element in Figure 4.4 with the straight segments $\overline{PO}$, $\overline{OS}$, and $\overline{OG}$ with O being the element's centroid, divide the triangular element into three equal polygons each 1/3 of the total area of the triangle. Assuming that the total amount of electrical charges inside it are distributed uniformly on the delimited volume by the prism of the triangular base and unitary height, according to a constant volumetric density of charge ρ (C/m³), we can write

$$Q_1^e = Q_2^e = Q_3^e = \rho \frac{\Delta}{3}$$

or, in matrix form

$$\begin{bmatrix} Q_1^e \\ Q_2^e \\ Q_3^e \end{bmatrix} = \begin{bmatrix} \rho\Delta/3 \\ \rho\Delta/3 \\ \rho\Delta/3 \end{bmatrix} \tag{4.18}$$

The column vector of the expression (4.18) is called vector of local actions, which is associated with the electrical charges, the sources of the electric field.

Finally, Maxwell's fourth equation on a closed surface involving node (i) can be applied as follows:

$$\sum_{e=1}^{NE} E_i^e = \sum_{e=1}^{NE} Q_i^e \quad i = 1, 2, \ldots, NN \tag{4.19}$$

where
NE is the total of triangular elements of the mesh
NN is its total number of nodes

It is highlighted that the terms from the sums in (4.19) will only have non-null values on the elements (e's) that admit node (i) as a vertex. This expression also generates a system of NN equations with NN unknowns, which are the electric potential of nodes, for which we can write

$$[C][V] = [Q] \tag{4.20}$$

As the matrix [C] is assembled from the elements' matrixes that are singular matrixes, [C] is also singular. This singularity is eliminated after the introduction of the boundary conditions of the problem, as it will be discussed later. The assembly of this system of equations is made in expeditious form, through a very simple algorithm, which will be discussed later in a simple application's exercise.

The resolution of the equations' system obtained after the introduction of the boundary conditions provides the electric potentials of all nodes in the domain. The knowledge of the electric potentials of nodes allows us to evaluate not only the electric field vector distribution inside the elements, but also other interesting quantities, such as capacitances, electrostatic potential energy, forces and torques of electrostatic nature, etc.

4.6 2D ELECTROKINETICS: STATIONARY CURRENTS FIELD

Another phenomenon governed by equations similar to those in electrostatics is electrokinetics, also called stationary currents field. In this study, perfect conductors excited by voltage sources are immersed in real conductors' media, resulting in an electric current flux through these medium. Situations like these occur in resistors, in ground meshes, and in other devices (Figure 4.8).

The starting point for this development is the continuity equation (or the law of conservation of charge) given by

$$\oint_{\Sigma} \vec{J} \cdot d\vec{S} = 0 \qquad (4.21)$$

where $\vec{J}$ is the current density vector (A/m²).

The current density vector and the electric field vector are related by the constitutive relation $\vec{J} = \sigma\vec{E}$, called Ohm's law, where σ is the electric conductivity of the medium measured in S/m.

As presented earlier, the relation between the electric field vector and the electric potential function is

$$\vec{E} = -\nabla V \qquad (4.22)$$

Table 4.1 shows a comparison between the electrostatics and electrokinetics equations, which makes our deductive process easier.

Thus, replacing appropriately ($\vec{D}$ by $\vec{J}$ and ε by σ and imposing $Q = 0$) on the electrostatics equations, we obtain the electrokinetics equations, and the electrokinetics equation (4.17) is written in matrix form as follows:

$$\begin{bmatrix} E_1^e \\ E_2^e \\ E_3^e \end{bmatrix} = \frac{\sigma}{4\Delta} \begin{pmatrix} b_1b_1 + c_1c_1 & b_1b_2 + c_1c_2 & b_1b_3 + c_1c_3 \\ b_2b_1 + c_2c_1 & b_2b_2 + c_2c_2 & b_2b_3 + c_2c_3 \\ b_3b_1 + c_3c_1 & b_3b_2 + c_3c_2 & b_3b_3 + c_3c_3 \end{pmatrix} \begin{bmatrix} V_1 \\ V_2 \\ V_3 \end{bmatrix} \qquad (4.23)$$

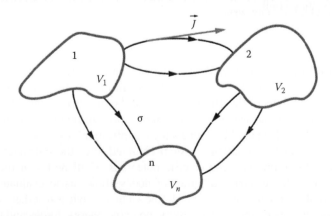

FIGURE 4.8 A typical problem in electrokinetics.

TABLE 4.1

Electrostatics/Electrokinetics Duality

Electrostatics	Electrokinetics
$\oint_{\Sigma} \vec{D} \cdot d\vec{S} = Q_i$	$\oint_{\Sigma} \vec{J} \cdot d\vec{S} = 0$
$\vec{E} = -\nabla V$	$\vec{E} = -\nabla V$
$\vec{D} = \varepsilon \vec{E}$	$\vec{J} = \sigma \vec{E}$
$\vec{D}\ (\text{C/m}^2)$	$\vec{J}\ (\text{A/m}^2)$
$\varepsilon\ \ (\text{F/m})$	$\sigma\ \ (\text{S/m})$

As the second member of the continuity equation (4.21) is null, there is no vector corresponding to (4.18), so that Equation 4.19 is reduced to

$$\sum_{e=1}^{NE} E_i^e = 0 \quad i = 1, 2, \ldots, NN \tag{4.24}$$

This can be expressed in matrix form as follows:

$$\left[G \right]\left[V \right] = \left[0 \right] \tag{4.25}$$

The similarity between the equation systems (4.20) and (4.25) is also undetermined. This indetermination is eliminated with the introduction of the boundary conditions of the problem as we will see later.

4.7 MAGNETOSTATICS

The mathematical formulation of magnetostatics, although different from those of electrostatics and electrokinetics, curiously leads to very similar results, as we will see.

Magnetostatics studies the magnetic field produced by DC electric currents. Figure 4.9 shows a typical magnetostatic two-dimensional problem, where the electric currents flow inside the conductors in the direction of the z-axis and the magnetic field produced is placed on the work plane (x, y *plane*).

In this case, the equation that governs the phenomenon is Maxwell's second equation, given by

$$\oint_{C} \vec{H} \cdot d\vec{l} = \int_{S} \vec{J} \cdot d\vec{S} \tag{4.26}$$

The idea is the same applied in the early items. The difference is that we need to choose a closed path in spite of a volume as we used in both electrostatic and electrokinetic studies due to the fact that the second Maxwell's equation involves a circulation of the magnetic field intensity different from the volume integration used in electrostatics and electrokinetics. In this case, the closed path that will be chosen

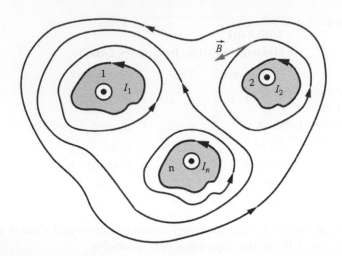

FIGURE 4.9   A typical problem in magnetostatics.

involving each node of the mesh is the same polygonal line linking the triangle's centroid to the center of each of its edges.

The circulation—the line integral—will be evaluated over all polygonal lines of the magnetic field vector, that is, the left term of Maxwell's second equation. The surface integration of the right term of (4.26) will be performed over all surfaces delimited by this closed line. Observe that the left term is the linkage current to the polygon.

The constitutive relation to be considered is the one that relates the magnetic field vector ($\vec{H}$) and the magnetic flux density field ($\vec{B}$):

$$\vec{H} = \nu\vec{B}$$

where $\nu = 1/\mu$ is the reluctivity of the medium (m/H).

As already discussed, the fact that $\nabla \cdot \vec{B} = 0$ allows to define the magnetic potential vector $\vec{A}$ leads to

$$\vec{B} = \nabla \times \vec{A} \tag{4.27}$$

To ensure uniqueness of the solution, the following must also be imposed

$$\nabla \cdot \vec{A} = 0$$

As already mentioned, in the case of 2D magnetostatics fields, the electric currents flow in the normal direction to the work plane (*perpendicular to the sheet*), so that

$$\vec{J} = J(x, y)\vec{u}_z$$

As $\vec{J}$ and $\vec{A}$ are aligned, we can write

$$\vec{A} = A(x, y)\vec{u}_z$$

Thus, Equation 4.27 is written as follows:

$$\vec{B} = \frac{\partial A}{\partial y} \vec{u}_x - \frac{\partial A}{\partial x} \vec{u}_y \tag{4.28}$$

As the components of the magnetic potential vector have the same continuity properties of the electric potential function, we can evaluate the value of component (z) of the magnetic potential vector in any point inside the element through the same interpolation function, that is,

$$A(x, y) = N_1 A_1 + N_2 A_2 + N_3 A_3 \tag{4.29}$$

where

$$N_i = \frac{1}{2\Delta}\left(a_i + b_i x + c_i y\right) \quad i = 1, 2, 3$$

Therefore, from (4.28), we get

$$B_x = \frac{\partial A}{\partial y} = \frac{1}{2\Delta}\left(c_1 A_1 + c_2 A_2 + c_3 A_3\right)$$

$$B_y = -\frac{\partial A}{\partial x} = -\frac{1}{2\Delta}\left(b_1 A_1 + b_2 A_2 + b_3 A_3\right)$$

Applying the constitutive relation $\vec{H} = v\vec{B}$, we get

$$H_x = vB_x = \frac{v}{2\Delta}\left(c_1 A_1 + c_2 A_2 + c_3 A_3\right)$$
$$\tag{4.30}$$
$$H_y = vB_y = -\frac{v}{2\Delta}\left(b_1 A_1 + b_2 A_2 + b_3 A_3\right)$$

Reporting to the generic element from Figure 4.10, the following line integrals are defined:

$$E_1^e = \int_{POS} \vec{H} \cdot d\vec{l} = H_x \Delta x - H_y \Delta y \tag{4.31}$$

This expression represents a section of the circulation of $\vec{H}$ on the polygonal line involving node (1), which belongs to the generic element (e). Similarly, the following integrals are defined:

$$E_2^e = \int_{GOP} \vec{H} \cdot d\vec{l} \quad \text{and} \quad E_3^e = \int_{SOG} \vec{H} \cdot d\vec{l}$$

which are sections of $\vec{H}$ circulations on the polygon line involving nodes (2) and (3), respectively.

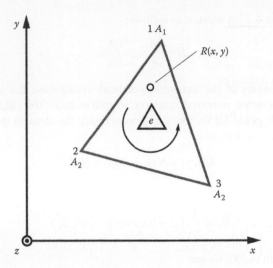

FIGURE 4.10   Generic element in magnetostatics.

Replacing H_x and H_y by its values expressed in (4.31) and observing that

$$\Delta x = x_s - x_p = \frac{1}{2}(x_3 - x_2) = \frac{c_1}{2}$$

$$\Delta y = y_p - y_s = \frac{1}{2}(y_2 - y_3) = \frac{b_1}{2}$$

results in

$$E_1^e = \frac{\nu}{4\Delta}\left[\left(b_1 b_1 + c_1 c_1\right)A_1 + \left(b_1 b_2 + c_1 c_2\right)A_2 + \left(b_1 b_3 + c_1 c_3\right)A_3\right] \qquad (4.32)$$

For the circulation section of $\vec{H}$ on the closed path that involves node (2) from e-ith element, we can write

$$E_2^e = \int_{GOP} \vec{H}\cdot d\vec{l} = H_x \Delta x + H_y \Delta y \qquad (4.33)$$

with

$$\Delta x = x_p - x_g = \frac{c_2}{2}$$

$$\Delta y = y_p - y_g = -\frac{b_2}{2}$$

Replacing H_x and H_y with their values expressed in (4.30) and Δx and Δy with their values from this expression, we obtain

$$E_2^e = \frac{v}{4\Delta}\left[\left(b_2b_1 + c_2c_1\right)A_1 + \left(b_2b_2 + c_2c_2\right)A_2 + \left(b_2b_3 + c_2c_3\right)A_3\right] \qquad (4.34)$$

For the circulation section $\vec{H}$ on the polygonal line that involves node (3) from e-ith element, by analog procedure we obtain

$$E_3^e = \frac{v}{4\Delta}\left[\left(b_3b_1 + c_3c_1\right)A_1 + \left(b_3b_2 + c_3c_2\right)A_2 + \left(b_3b_3 + c_3c_3\right)A_3\right] \qquad (4.35)$$

The results obtained in (4.32), (4.34), and (4.35) can be represented in matrix form as follows:

$$\begin{bmatrix} E_1^e \\ E_2^e \\ E_3^e \end{bmatrix} = \frac{v}{4\Delta}\begin{pmatrix} b_1b_1 + c_1c_1 & b_1b_2 + c_1c_2 & b_1b_3 + c_1c_3 \\ b_2b_1 + c_2c_1 & b_2b_2 + c_2c_2 & b_2b_3 + c_2c_3 \\ b_3b_1 + c_3c_1 & b_3b_2 + c_3c_2 & b_3b_3 + c_3c_3 \end{pmatrix}\begin{bmatrix} A_1 \\ A_2 \\ A_3 \end{bmatrix} \qquad (4.36)$$

The only difference between the magnetostatics matrix element and those of electrostatics and electrokinetics is the material properties constant. This is very convenient because only one computational procedure is enough for simulating any kind of 2D static electromagnetic phenomenon.

The left term of Maxwell's second equation corresponds to the linkage current by the polygon line.

The total linkage current that crosses the surface of the polygonal line is composed of several portions of currents issued from all elements that have their node considered as vertex. Thus, from each element, we extract three portions that will contribute to three different polygonal lines, one around each node, and we can write

$$I_1^e = J\frac{\Delta}{3}$$

which represents the portion of the total current of the generic element (e), which is linkaged with the polygonal line that involves node (1), because the centroid with the points at the middle of the triangle edge including its vertex divides the surface of the element into three identical surfaces. Likewise, we identify the other portions as follows:

$$I_2^e = I_3^e = J\frac{\Delta}{3}$$

This is written in matrix form as follows:

$$\begin{bmatrix} I_1^e \\ I_2^e \\ I_3^e \end{bmatrix} = \begin{bmatrix} J\Delta/3 \\ J\Delta/3 \\ J\Delta/3 \end{bmatrix} \qquad (4.37)$$

Finally, the application of Maxwell's second equation on a polygonal line involving node (i) can be written as follows:

$$\sum_{e=1}^{NE} E_i^e = \sum_{e=1}^{NE} I_i^e \quad i = 1, 2, \ldots, NN \qquad (4.38)$$

Where

NE is the total number of elements from the mesh
NN is its total number of nodes

It is noted again that the non-null terms of the indicated sums in (4.38) are the ones that admit node (i) as a vertex. As in (4.20), this expression also generates a system of *NN* equations with *NN* unknowns, which represent the components (z) of the magnetic potential vector of the node. Hence, we can write

$$[S][A] = [I] \qquad (4.39)$$

The matrix [S] from magnetostatic study also has null determinant due to the similarity of both matrix [C] from electrostatics and [G] from electrokinetics, so the system (4.39) is also undetermined.

We suggest that the elements from the matrixes of the three types of studies be practically identical, only differing in physical property, so that a code written for electrostatics can also be used for the other two.

4.8 ASSEMBLING OF THE GLOBAL SYSTEM OF EQUATIONS

We are already aware of the assembly algorithm of the global system of equations. Due to the characteristics of the presented methodology, the assembly of this global system of equations is made based on a very simple algorithm, which we will describe in this section.

Each element of the domain is treated separately and we associate two numbering methods to its vertexes (or nodes). The first one, called "global" numbering, involves arbitrary sequential numbering of all the nodes from the mesh of finite elements, which goes from (1) to (*NN*). There is no defined criterion for this numbering, which is done randomly. The second numbering is called "local" numbering, which consists in numbering the vertexes of each element from (1) to (3) on counterclockwise, as shown in Figure 4.11.

The association between these numbering methods is given in Table 4.2.

The knowledge of the physical properties of the medium of the element; the coordinates from their vertexes and sources (*current density charges or electrical charges*) two matrixes are assembled depending on the type of study:

1. The matrix from the element (*C*, *G*, or *S*) of dimension (3 × 3)
2. The vector of the actions (*Q* or *I*) of dimension (3 × 1)

Global matrix: In the global matrix, we begin assembling a square matrix of order *NN* constituted by null elements.

The matrix of the element is then transported to the global matrix, following the association between "local" and "global" numbering. Thus, element (1, 1) from

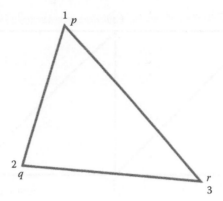

FIGURE 4.11 Global numbering and local numbering.

TABLE 4.2
"Local" and "Global" Numbering

Local	1	2	3
Global	p	q	r

---

the matrix of the element is summed to element (p, p) from the global matrix, element $(1, 2)$ is summed to element (p, q) from the global matrix, and so on. This process is repeated for all elements of the mesh [23,62].

Vector of actions: The assembly of the vector from global actions is similar. The process starts with assembling a null element's column vector with (NN) lines. Each element will contribute three elements to this column vector, so that line (1) from the element's actions vector is added to line (p) from the global actions vector, line (2) to line (q), and line (3) to line (r), repeating this process for all the elements of the mesh.

As a result, the global matrix is both symmetric and sparse. Due to the characteristics of the global matrix—symmetry and sparsity—several compact procedure routines are available for optimizing the management of the computing memory.

For a better understanding of this operation, let us see how the global system of equations from the simple domain from Figure 4.12 is assembled, which is constituted by four elements (numbered 1–4) and six nodes (numbered 1–6).

The association between the "local" and "global" numbering of all the elements from the domain is shown in Table 4.3.

Representing the values of the matrixes of element (1) by (#), we obtain two matrixes associated with this element as follows:

$$
\begin{pmatrix} \# & \# & \# \\ \# & \# & \# \\ \# & \# & \# \end{pmatrix} \text{Matrix of element 1} \quad \begin{pmatrix} \# \\ \# \\ \# \end{pmatrix} \text{Action vector of element 1}
$$

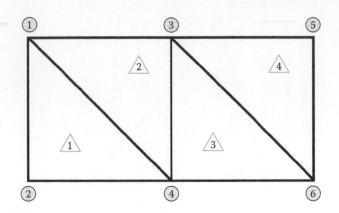

FIGURE 4.12 Domain under study.

TABLE 4.3
Correspondences between Numberings

Local Numbering	1	2	3
Element (1)	1	2	4
Element (2)	1	4	3
Element (3)	3	4	6
Element (4)	3	6	5

For the values of the matrixes from element (2) we will use the symbol (□) so that for this element we obtain the following matrixes:

$$
\begin{pmatrix} \square & \square & \square \\ \square & \square & \square \\ \square & \square & \square \end{pmatrix} \text{Matrix of element 2} \quad \begin{pmatrix} \square \\ \square \\ \square \end{pmatrix} \text{Action vector of element 2}
$$

Same for the elements (3) and (4):

$$
\begin{pmatrix} \Delta & \Delta & \Delta \\ \Delta & \Delta & \Delta \\ \Delta & \Delta & \Delta \end{pmatrix} \text{Matrix of element 3} \quad \begin{pmatrix} \Delta \\ \Delta \\ \Delta \end{pmatrix} \text{Action vector of element 3}
$$

$$
\begin{pmatrix} * & * & * \\ * & * & * \\ * & * & * \end{pmatrix} \text{Matrix of element 4} \quad \begin{pmatrix} * \\ * \\ * \end{pmatrix} \text{Action vector of element 4}
$$

The global system of the resulting equations is given in Figure 4.13.

$$
\begin{pmatrix}
\#\square & \# & \square & \#\square & 0 & 0 \\
\# & \# & 0 & \# & 0 & 0 \\
\square & 0 & {}^*\square\Delta & \Delta\square & * & \Delta^* \\
\#\square & \# & \square\Delta & \Delta\#\square & 0 & \Delta^* \\
0 & 0 & * & 0 & * & 0 \\
0 & 0 & {}^*\Delta & \Delta & * & \Delta^*
\end{pmatrix}
\cdot
\begin{pmatrix}
\theta_1 \\ \theta_2 \\ \theta_3 \\ \theta_4 \\ \theta_5 \\ \theta_6
\end{pmatrix}
=
\begin{pmatrix}
\#\square \\ \# \\ \square\Delta^* \\ \#\square\Delta \\ * \\ \Delta^*
\end{pmatrix}
$$

$$\qquad\qquad\text{(a)}\qquad\qquad\qquad\qquad\text{(b)}\quad\text{(c)}$$

FIGURE 4.13 Global system of equations. (a) Global matrix (b) vector of unknowns (*A* or *V*) (c) vector of global actions.

Where are we?: Until this moment, we learnt to assemble an undetermined system of equations resulting from the transformation of an electromagnetic field problem, described by Maxwell's equations, in an algebraic equations system.

In general, for any type of study, the global system of equations is as follows:

$$[K]\cdot[\theta]=[P] \tag{4.40}$$

Depending on the type of study $[K]$ can be $[C]$, $[G]$, or $[S]$; $[\theta]$ equal to $[V]$ or $[A]$, and finally $[P]$ equal to $[Q]$, $[0]$, or $[I]$.

As in all problems involving integrals and/or differential equations, the final solution is only obtained after the introduction of the boundary conditions. For the 2D static electromagnetics phenomenon, these boundary conditions are the known values of the state variable (V or A) in some boundary's nodes of the mesh.

4.9 INTRODUCTION OF THE BOUNDARY CONDITIONS

It is convenient at this point to adopt a little formalism in our discussion for making the introduction of the boundary conditions on the global system of equations easier.

This was addressed during the definition process of the problem, which includes the definition of the study domain.

Thus, in a two-dimensional case, the domain under study, which we will call Δ, is limited by a boundary called S. This S boundary is divided into two parts, S_1 and S_2 with $S = S_1 + S_2$, which can be defined as follows:

S_1: Part of the boundary where the variable of the state (potentials in case of electromagnetism) is known, in other words, $\theta = \theta_0$ over S_1.

S_2: Part of the boundary where the normal derivative from the state variable is known, in other words, $\partial\theta/\partial n = g$ over S_2.

The first type of boundary conditions is called *Dirichlet's conditions*. In electromagnetism studies, the Dirichlet conditions are found under the following conditions:

Electrostatics and electrokinetics: In perfect conductors submitted to electric potentials imposed by voltage sources, $V = V_0$ on conductors, with V_0 equal to the potential imposed on the conductor.

Magnetostatics: In the part of the boundary located in the region where the magnetic flux density is null, the (z component of) magnetic potential vector is also null, that is, $A = 0$ in this part of the boundary.

The second type of boundary conditions is called *Neumann's conditions*. In electromagnetism studies, Neumann's conditions are found under the following conditions:

Electrostatics and electrokinetics: On the part of the domain boundary where electric field lines are tangent to the boundary, that is, the normal component of $\vec{E}$ or $E_N = \partial V / \partial n = 0$ over S_2.

Magnetostatics: On the part of the domain boundary where magnetic field lines are normal to the boundary, that is, the tangential component of $\vec{B}$ or $B_T = \partial A / \partial n = 0$ over S_2.

4.9.1 INTRODUCTION OF DIRICHLET CONDITIONS

The division of the domain on finite elements will allocate some nodes (vertexes from elements) on the domain limits, the electric or magnetic potentials of which are known. As the equations of these nodes are part of a global equations system, we need to modify the global system in order to obtain, after its solution, the electric or magnetic potential of the nodes located at this part of the boundary result on the imposed values.

Now, suppose that (M) is a node of the finite elements mesh whose potential is given by the value $(\theta_M = \phi)$. To obtain $\theta_M = \phi$, we should just modify the global matrix and the actions vector as follows:

$$K_{MM} = 1 \quad \text{and} \quad K_{MJ} = 0 \quad \text{for } j \neq m \quad \text{and} \quad P_M = \phi$$

These procedures eliminate the symmetry of the global equations system, which is not convenient. For recovering this symmetry, we impose the following:

$$P_J = P_J - K_{JM}\phi \quad \text{and} \quad K_{JM} = 0 \quad \text{for } j \neq m$$

As an example, let us suppose that we know the potential of node 4. Let ϕ be its value. Applying the proposed algorithm in the equation's system of Figure 4.13, the new configuration of the linear system in Figure 4.14 becomes

4.9.2 INTRODUCTION OF NEUMANN CONDITIONS

Neumann conditions are naturally imposed when the domain is defined, ensuring that the part of S_2 from domain's boundary is such that (Figure 4.15)

$$\begin{pmatrix} \#\square & \# & \square & 0 & 0 & 0 \\ \# & \# & 0 & 0 & 0 & 0 \\ \square & 0 & ^*\square\Delta & 0 & ^* & \Delta^* \\ 0 & 0 & 0 & 1 & 0 & 0 \\ 0 & 0 & ^* & 0 & ^* & 0 \\ 0 & 0 & ^*\Delta & 0 & ^* & \Delta^* \end{pmatrix} \cdot \begin{pmatrix} \theta_1 \\ \theta_2 \\ \theta_3 \\ \theta_4 \\ \theta_5 \\ \theta_6 \end{pmatrix} = \begin{pmatrix} \#\square - k_{14}\Phi \\ \# - k_{24}\Phi \\ \square\Delta^* - k_{34}\Phi \\ \Phi \\ ^* - k_{54}\Phi \\ \Delta^* - k_{64}\Phi \end{pmatrix}$$

FIGURE 4.14   Global system of equations after introduction of boundary condition.

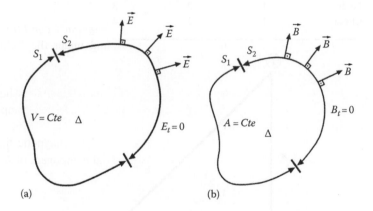

FIGURE 4.15   Boundary conditions in electromagnetism: (a) Electrostatics and electrokinetics. (b) Magnetostatics.

1. The electric field is tangent to S_2, ensuring that

$$E_i^e = \int_{S_2} \vec{D} \cdot d\vec{S} = 0$$

   in electrostatics or electrokinetics.
2. The magnetic field is normal to S_2, ensuring that

$$E_i^e = \int_{S_2} \vec{H} \cdot d\vec{l} = 0$$

   in magnetostatics.

Obs.: Neumann conditions are found when the symmetry of the domain is explored for reducing its dimensions. A common example is the study of electric machines that have several poles, which in FEM is considered only a pole for performing the study.

Example 4.1

Figure 4.16 shows a longitudinal cross section of a resistor with 2 m length, with both width and height unitary. The conductivity of the medium is 2 S/m. On the lateral faces a voltage of 100 V is applied. Calculate, applying the presented methodology, the electric potential distribution on the resistor and the electric field on all mesh elements.

1. *Evaluation of elements matrixes*
   Element (1)

   Correspondence between local and global numbering

Local	1	2	3
Global	1	2	4

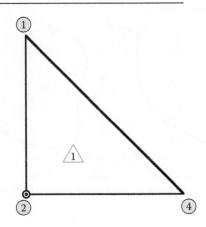

$$x_1 = 0; \; x_2 = 0; \; x_3 = 1$$

$$y_1 = 1; \; y_2 = 0; \; y_3 = 1$$

$$b_1 = y_2 - y_3; \; b_1 = 0 \quad c_1 = x_3 - x_2; \; c_1 = 1$$

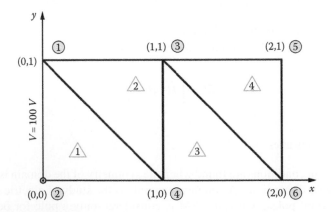

FIGURE 4.16   Resistor.

$$b_2 = y_3 - y_1; \; b_2 = -1 \quad c_2 = x_1 - x_3; \; c_2 = -1$$

$$b_3 = y_1 - y_2; \; b_3 = 1 \quad c_3 = x_2 - x_1; \; c_1 = 0$$

$$\Delta = (b_1 c_2 - b_2 c_1)/2; \; \Delta = 0,5$$

Matrix of the Element (1)

$$\begin{pmatrix} 1 & -1 & 0 \\ -1 & 2 & -1 \\ 0 & -1 & 1 \end{pmatrix}$$

Following the same procedure for the other elements, we obtain

Element (2)

$$\begin{pmatrix} 1 & 0 & -1 \\ 0 & 1 & -1 \\ -1 & -1 & 2 \end{pmatrix}$$

Element (3)

$$\begin{pmatrix} 1 & -1 & 0 \\ -1 & 2 & -1 \\ 0 & -1 & 1 \end{pmatrix}$$

Element (4)

$$\begin{pmatrix} 1 & 0 & -1 \\ 0 & 1 & -1 \\ -1 & -1 & 2 \end{pmatrix}$$

Obs.: In electrokinetics, the actions' vector is null.

2. *Assembling the Global Matrix*

$$\begin{pmatrix} 2 & -1 & -1 & 0 & 0 & 0 \\ -1 & 2 & 0 & -1 & 0 & 0 \\ -1 & 0 & 4 & -2 & -1 & 0 \\ 0 & -1 & -2 & 4 & -1 & -1 \\ 0 & 0 & -1 & 0 & 2 & -1 \\ 0 & 0 & 0 & -1 & -1 & 2 \end{pmatrix} \begin{pmatrix} V_1 \\ V_2 \\ V_3 \\ V_4 \\ V_5 \\ V_6 \end{pmatrix} = \begin{pmatrix} 0 \\ 0 \\ 0 \\ 0 \\ 0 \\ 0 \end{pmatrix}$$

3. *Introduction of the Boundary Conditions*

Given potentials: $V_1 = 100$ (V); $V_2 = 100$ (V); $V_5 = 0$ (V) e $V_6 = 0$ (V).
The modified equations system after the introduction of the boundary conditions is

$$\begin{pmatrix} 1 & 0 & 0 & 0 & 0 & 0 \\ 0 & 1 & 0 & 0 & 0 & 0 \\ 0 & 0 & 4 & -2 & 0 & 0 \\ 0 & 0 & -2 & 4 & 0 & 0 \\ 0 & 0 & 0 & 0 & 1 & 0 \\ 0 & 0 & 0 & 0 & 0 & 1 \end{pmatrix} \begin{pmatrix} V_1 \\ V_2 \\ V_3 \\ V_4 \\ V_5 \\ V_6 \end{pmatrix} = \begin{pmatrix} 100 \\ 100 \\ 100 \\ 100 \\ 0 \\ 0 \end{pmatrix}$$

The solution of this equation system is

$V_1 = 100$ (V)	$V_3 = 50$ (V)	$V_5 = 0$ (V)
$V_2 = 100$ (V)	$V_4 = 50$ (V)	$V_6 = 0$ (V)

Observe that no procedure is necessary for imposing Neumann's conditions that are presented on the boundary domain over the nodes 1, 3, and 5 or 2, 4, and 6.

4. *Evaluation of the electric field*

From (4.8), for the element (1) we obtain

$$E_{x1} = -[(0) \times 100 + (-1) \times 100 + (1) \times 50]; \; E_{x1} = 50 \text{ V/m}$$

$$E_{y1} = -[(1) \times 100 + (-1) \times 100 + (0) \times 50]; \; E_{y1} = 0 \text{ V/m (as we expected!!)}$$

Elements 2, 3, and 4 will also have the same values.

It is important to observe that the results obtained are exact; however, for real-world situations of a more complex domain, such accuracy is never reached.

Example 4.2

Figure 4.17 shows a square domain divided into two triangular elements with its corners numbered from 1 to 4.

Element (1) of vertices 1, 2, and 4 is passed through by a current density of $J_1 = 9000$ A/m². In element (2) of vertices 2, 3, and 4, the current density is $J_2 = 15,000$ A/m², both directed according to the z axis as shown in Figure 4.17. The magnetic reluctivity of the element is $v = 10,000$ m/H.

Let us determine the magnetic flux density inside each triangle, knowing that the magnetic potential vector in node 3 is null ($A_3 = 0$).

The following table shows the incidence between the nodes and the elements.

Element	1	2	3	J (A/m²)	v (m/H)
1	4	1	2	9,000	10,000
2	3	4	2	15,000	10,000

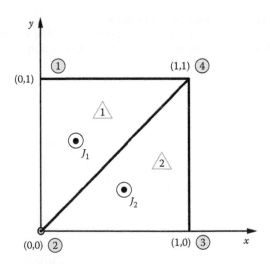

FIGURE 4.17   Domain under study—Dimensions in meters.

From the Chapter 1, we have

Element	b_1	b_2	b_3	c_1	c_2	c_3
1	1	-1	0	0	1	-1
2	1	0	-1	-1	1	0

From (4.36) and (4.37), we assemble the matrixes of the element as follows:

$$S^1 = \begin{pmatrix} 5000 & -5000 & 0 \\ -5000 & 10000 & -5000 \\ 0 & -5000 & 5000 \end{pmatrix} \quad l^1 = \begin{pmatrix} 1500 \\ 1500 \\ 1500 \end{pmatrix}$$

$$S^2 = \begin{pmatrix} 10000 & -5000 & -5000 \\ -5000 & 5000 & 0 \\ -5000 & 0 & 5000 \end{pmatrix} \quad l^2 = \begin{pmatrix} 2500 \\ 2500 \\ 2500 \end{pmatrix}$$

Following the proposed procedure, the next step is to assemble the global matrices:

$$\begin{pmatrix} 10000 & -5000 & 0 & -5000 \\ -5000 & 10000 & -5000 & 0 \\ 0 & -5000 & 10000 & -5000 \\ -5000 & 0 & -5000 & 10000 \end{pmatrix} \cdot \begin{pmatrix} A_1 \\ A_2 \\ A_3 \\ A_4 \end{pmatrix} = \begin{pmatrix} 1500 \\ 4000 \\ 2500 \\ 4000 \end{pmatrix}$$

Introducing the Dirichlet boundary conditions imposing $A_3 = 0$, the final system's equation becomes

$$\begin{pmatrix} 10000 & -5000 & 0 & -5000 \\ -5000 & 10000 & 0 & 0 \\ 0 & 0 & 1 & 0 \\ -5000 & 0 & 0 & 10000 \end{pmatrix} \cdot \begin{pmatrix} A_1 \\ A_2 \\ A_3 \\ A_4 \end{pmatrix} = \begin{pmatrix} 1500 \\ 4000 \\ 0 \\ 4000 \end{pmatrix}$$

Solving this system gives

$A_1 = 0.0769$ Wb/m; $A_2 = 0.2923$ Wb/m; $A_3 = 0$ Wb/m, and $A_4 = 0.4385$ Wb/m

With the magnetic potential vector, it is possible to evaluate the magnetic flux density in each element:

Element	B_x (Wb/m²)	B_y (Wb/m²)	B (Wb/m²)
1	-0.2154	0.3616	0.4209
2	0.4385	-0.2923	0.5270

4.10 INTRODUCTION OF SPECIAL BOUNDARY CONDITIONS

4.10.1 PERIODICITY CONDITIONS

The geometry of a large number of electromagnetic devices has repetitive sections, as it is the case with multipolar rotating electric machines. In such situations, analysis by FEM can be elaborated from the analysis of only a portion of this device, which is very convenient, because we can then obtain a sensible reduction of the dimensions of the domain under study.

Besides giving us a bigger picture of the problem, this procedure allows exploring the potentiality of the computational resources.

This type of boundary conditions can be understood better with the help of the geometry of multipolar synchronous machine cross sections shown in Figure 4.18.

Due to the symmetry, it is verified that the magnetic flux density at points (1) and (3) are the same, so that the (z component from) magnetic potential vector (A) will also be the same. In that case, we can write

$$A(1) = A(3)$$

This kind of boundary condition is called "cyclic condition." Therefore, it is not necessary to consider the entire domain for performing the complete analysis of the machine, because what happens in this part repeats in the other identical parts of the remaining domain. For this reason, we can constrain the domain under study to the region comprised by the polygon $XX'Z'ZX$, that is, two polar pitches.

Otherwise, the magnetic flux density in points (1) and (2) have also the same amplitude, but with different senses related to the polar section. In that case, we can write

$$A(1) = -A(2)$$

This kind of boundary condition is called "anticyclic" condition. In these cases, the domain under study will be restricted to the region comprised by the polygon $XX'Y'YZ$, that is, a polar pitch.

When these types of boundary conditions are explored, the number of nodes present on the symmetry lines should be the same and spaced in the same way.

The introduction of these boundary conditions implies the reduction of the equations system, because a series of unknown parts of the problem are the same

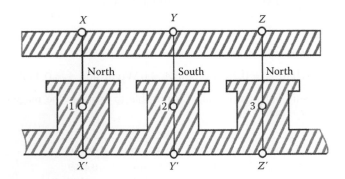

FIGURE 4.18   Multipolar synchronous machine.

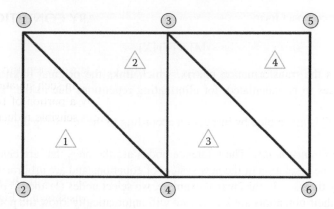

FIGURE 4.19   Domain under study.

between them, as in the case of the "cyclic" condition, or opposed between them, as in the case of the "anticyclic" condition.

For illustrating how these types of boundary conditions are imposed in the global system, let us look back at the domain from Example 4.1.

Supposing that the boundary potentials (1) and (2) are opposite to the boundary potentials (5) and (6), that is, in these lines we will impose the "anticyclic" condition, we get (Figure 4.19)

$$A(1) = -A(5)$$

$$A(2) = -A(6)$$

Therefore, only the variables in nodes 1, 2, 3, and 4 need to be calculated.

The introduction of these conditions implies on the modification of the size of the original system from 6×6 to a system of order 4×4.

X_1, X_2, X_3, and X_4 being the new unknowns from the transformed system, the relations between these and the unknowns from the original system are such that

$$A_1 = X_1; \; A_2 = X_2; \; A_3 = X_3; \; A_4 = X_4; \; A_5 = -X_1; \; A_6 = -X_2$$

These relations can be represented in matrix form as follows:

$$
\begin{bmatrix} A_1 \\ A_2 \\ A_3 \\ A_4 \\ A_5 \\ A_6 \end{bmatrix}
=
\begin{pmatrix} 1 & 0 & 0 & 0 \\ 0 & 1 & 0 & 0 \\ 0 & 0 & 1 & 0 \\ 0 & 0 & 0 & 1 \\ -1 & 0 & 0 & 0 \\ 0 & -1 & 0 & 0 \end{pmatrix}
\begin{bmatrix} X_1 \\ X_2 \\ X_3 \\ X_4 \end{bmatrix}
$$

or under the compact form

$$[A] = [T] \cdot [X] \tag{4.41}$$

where $[T]$ is the transformation matrix, which links the original unknowns with the news ones to be calculated, for eliminating repetitions due to the "anticyclic" condition.

Matrix $[T]$ is assembled by inspection according to the following algorithm:

1. *For reference nodes*: The reference nodes are the ones that are calculated using the resolution of the new system of equations and are arbitrarily chosen by the user. In the current example, we select nodes (1) and (2) because once their potentials are known, we will automatically know the potentials of nodes (5) and (6).
   For these nodes, we write

$$T_{ij} = \begin{cases} 1 & \text{if } A_i = X_j \\ 0 & \text{if } A_i \neq X_j \end{cases}$$

2. For the nodes which will be eliminated

$$T_{ij} = \begin{cases} 1 & \text{if } A_i = X_j \text{ for cyclic condition} \\ -1 & \text{if } A_i = -X_j \text{ for anti-cyclic condition} \\ 0 & \text{if } A_i \neq X_j \end{cases}$$

In the case of magnetostatics analysis, the original system without the imposition of the boundary conditions is like $[S] \cdot [A] = [I]$, which is obtained according the proposed procedure. The order of this system is NN, that is, the number of nodes from the domain.

Replacing the original unknowns vector by its indicated transformation in (4.41), the original system is rewritten as follows:

$$[S] [T] [X] = [I]$$

Multiplying both sides by the transposed from transformation matrix results in

$$[T]^{\mathrm{T}}[S] [T] [X] = [T]^{\mathrm{T}}[I] \tag{4.42}$$

This results in a new system of equations to be solved:

$$[H] [X] = [B] \tag{4.43}$$

where

$$[H] = [T]^{\mathrm{T}}[S][T] \quad \text{and} \quad [B] = [T]^{\mathrm{T}}[I]$$

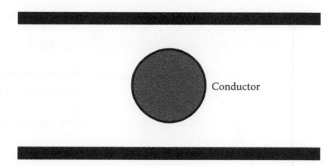

FIGURE 4.20 Capacitor with an internal conductor.

The order of this new system is $(NN - NP/2)$, where NP is the number of nodes with periodicity condition. The computational implementation of this type of boundary conditions can be elaborated without the assembly of the transformation matrix, with just a careful manipulation of the lines and columns of the matrix $[S]$.

4.10.2 FLOATING BOUNDARY CONDITION

This type of boundary condition is common in electrostatics and electrokinetics studies. For illustrating its application, suppose the plan capacitor shown in Figure 4.20, which contains on its interior a metal body placed between two plaques.

It is clear that the external surface of the internal conductor is an equipotential surface, the potential of which has an unknown value. Therefore, all the nodes placed in this surface have the same electric potential, and so only the potential of a single node from the conductor's surface needs to be calculated.

The imposition of this type of boundary condition is very similar to the imposition of the "cyclic" condition. However, the user must choose only a single reference node, and through the transformation matrix $[T]$ the condition of the same potential of all nodes of the conductor surface is imposed.

4.11 NONLINEAR MAGNETOSTATICS

The presence of ferromagnetic materials on electric devices is an additional difficulty that numerical methods have to face due to the nonlinearity of the magnetic reluctivity. It is a fact that the magnetic reluctivity of ferromagnetic materials depends on the magnetic flux density present on the structure, that is, $\nu = \nu(B)$. The evidence of this fact is centered on the nonlinear characteristic of the typical magnetization curve of ferromagnetic materials, as shown in Figure 4.21.

This problem is overcome through an iterative process, which estimates an initial value for the magnetic reluctivity, which is corrected step by step until an accuracy criterion is reached on the convergence process.

There are several iterative processes available; however, let us focus on the Newton–Raphson method, because it is the most disseminated and highly efficient in the treatment of electromagnetic problems.

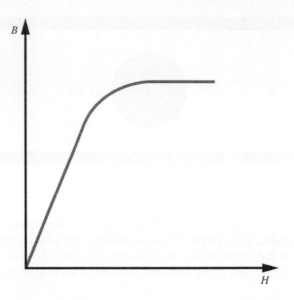

FIGURE 4.21 Typical magnetization curve of ferromagnetic materials.

The most frequent methodology used in this iterative process begins by obtaining the characteristic $\nu = \nu(B^2)$. The reason for the selection of the reluctivity representation (ν) based on (B^2) lies in the fact that it is necessary to evaluate it in each step of the process. Indeed, for evaluating the magnitude of the magnetic flux density (B), the value of B^2 should be calculated before; with this procedure, we save processing time.

The characteristic $\nu = \nu(B^2)$ is obtained directly from the magnetization curve. The process starts by choosing a sufficient number of representative points of the saturation curve (H_i, B_i) for evaluating the relation $\nu_i = H_i/B_i$ and graphically representing the curve $\nu_i = \nu_i(B_i^2)$.

The characteristic $\nu = \nu(B^2)$ extracted from the magnetization curve of the ferromagnetic material has the aspect shown in Figure 4.22.

Once this curve obtained, we only need to obtain a mathematical function for representing it. Many alternatives are presented [4]. One of them consists in dividing it in sections and adjusting for each section a kind of squared function, such as

$$\nu = \frac{B^2 - B_i^2}{B_j^2 - B_i^2} \tag{4.44}$$

From our experience, we suggest choosing 5–8 sections with most of them located near the knee of the saturation characteristics.

Another alternative, proposed by Hoole [6], consists in searching the coefficients, which help obtain the curve through the function

$$\nu = k_1 e^{k_2 x} + k_3 \quad \text{with } x = B^2 \tag{4.45}$$

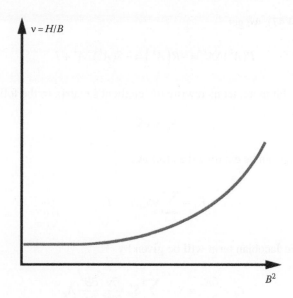

FIGURE 4.22 Characteristic $\nu = \nu(B^2)$.

The coefficients of (4.45) are obtained from fitting the curve's algorithm, for example, using the least squares method. This approach has the inconvenience of high magnetic density flux where it can produce oscillations on the solution while the Newton–Raphson method solves the system.

In magnetostatics, the solved equations system is the one presented in (4.39), that is

$$[S][A] = [I]$$

As the Newton–Raphson method is an iterative algorithm, no exact solution of the system is obtained, and it always results in an approximation residue on each iteration as follows:

$$R = SA - I \tag{4.46}$$

Expanding this residue by the Taylor's series and truncating the terms of superior order to one and also imposing that on the iteration $K + 1$ this residue is null, we obtain

$$R(A^{k+1}) = R(A^k) + \frac{\partial R(A^k)}{\partial A} \Delta A^k = 0 \tag{4.47}$$

with

$$\Delta A^k = A^{k+1} - A^k$$

The matrix

$$P(A^k) = \frac{\partial R(A^k)}{\partial A}$$

is called the Jacobian matrix.

Reordering (4.47), we get

$$P(A^k)\Delta A^k = -R\left(A^k\right) = -S(A^k) \cdot A^k + I \tag{4.48}$$

Detailing a little bit more, let us rewrite the element's matrix in the following way:

$$S_{ij} = \nu S'_{ij}$$

so that the residue can be expressed as follows:

$$R_j = \sum_{k=1}^{3} \nu S'_{kj} A^k - I_j$$

Thus, the generic Jacobian term will be given by

$$P_{ij} = \frac{\partial R_i}{\partial A_j} = S_{ij} + \sum_{k=1}^{3} S'_{ik} \frac{\partial \nu}{\partial B^2} \frac{\partial B^2}{\partial A_j} A_k \tag{4.49}$$

Remembering that

$$B_x = \frac{1}{2\Delta} \sum_{k=1}^{3} c_k A_k$$

and

$$B_y = -\frac{1}{2\Delta} \sum_{k=1}^{3} b_k A_k$$

we get

$$B^2 = B_x^2 + B_y^2 = \sum_{k=1}^{3} \sum_{l=1}^{3} \frac{S'_{kl}}{\Delta} A_k A_l \tag{4.50}$$

Deriving (4.50) related to A, we obtain

$$\frac{\partial B^2}{\partial A_j} = \frac{2}{\Delta} \sum_{k=1}^{3} S'_{jl} A_l \tag{4.51}$$

Replacing this result in (4.49) gives

$$P_{ij} = S_{ij} + \frac{2}{\Delta} \sum_{k=1}^{3} \sum_{l=1}^{3} S'_{ik} S'_{jl} \frac{\partial \nu}{\partial B^2} A_k A_l \tag{4.52}$$

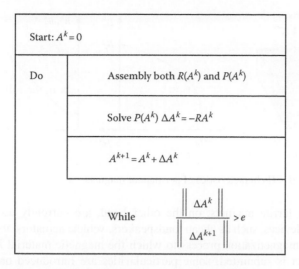

FIGURE 4.23   Structured algorithm for the Newton–Raphson method.

The solution of (4.48) is obtained through an iterative process where both the general terms, $P(A^k)$ and $R(A^k)$, are obtained by the expressions presented.

The system's convergence is tested at each step using

$$\frac{\left\| \Delta A^k \right\|}{\left\| \Delta A^{k+1} \right\|} \leq \varepsilon \tag{4.53}$$

where ε is the preestablished tolerable error value.

Normally, choosing $\varepsilon = 10^{-6}$ the number of iterations necessary for the convergence is about 10.

A flow chart for this procedure is represented in Figure 4.23.

4.12   MAGNETOSTATICS WITH PERMANENT MAGNETS AND ANISOTROPIC MEDIA

Several electromagnetic devices use permanent magnets in the production of necessary magnetic fields for their operations. The advantage of these permanent magnets lies in the fact that we have no necessity for voltage sources to feed the coils producing the magnetic field, contributing to a reduction in the device's dimensions and to improving its efficiency.

In modern electric machines, mainly those used for a fine control of the speed, the most used magnets are rare-earth magnets, such as samarium–cobalt and neodymium–iron–boron magnets. These products, which require high technology for their production, present a considerable energy density (maximum product $B \times H$), which attributes to design alternatives never made available before by ordinary permanent magnets, such as alnico, ferrite, etc.

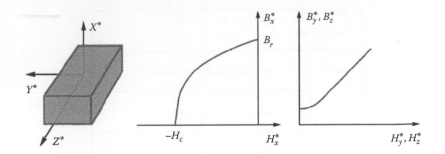

FIGURE 4.24 $B \times H$ characteristics of permanent magnets.

Alnico and ferrite magnets, on the other hand, are currently used in not-so-sophisticated devices, such as cheap loudspeakers, vehicle actuators, toys, etc.

Facing the magnetization process to which the magnetic material from the permanent magnet is submitted, some particularities are introduced on its physical behavior, which were not treated when we studied magnetostatic modeling.

A detailing of how this magnetization process can attribute its particular properties to the magnet is presented in Abe [7], and it will be reproduced herein.

A pastille of a permanent magnet submitted to tests through a "hysteresimeter," with the aim of raising its magnetic characteristics ($B \times H$ curves), presented the results shown in Figure 4.24.

An analysis of the data in this figure shows that the x^* direction, called the magnetization axis of the permanent magnet, is the direction in which the magnetic domains are preferentially aligned because the magnetic field is maximum in this direction.

In the y^* and z^* directions, the magnetization curves are—practically—straight. This property enable us to consider this evidence in our mathematical formulation and using it during the design of electrical devices. The angular coefficient of this straight curve is equal to the angular coefficient of the straight tangent to the magnetization curve of the magnet according to the x^* direction by point $(0, B_r)$. This angular coefficient is called the recoil permeability (μ_{rec}) of the magnet.

A mathematical treatment that simplifies the analysis of the permanent magnet is the one developed by Jordão [9], where a permanent magnet is replaced by a fictitious coil, having an MMF (magneto-motive force) that produces an internal magnetic field equal to the coercive field (H_c), as shown in Figure 4.25.

The replacement of permanent magnet by a fictitious coil that imposes an MMF ($H_c l_i$), where l_i is the length of the magnet, Implies that the medium inside the coil has the magnetic permeability different in both aligned and orthogonal directions with the magnetic coercive field. In other words, in this region, the magnetic material is anisotropic. Thus, we can write

$$H_{x^*} = v_1 B_{x^*}$$

where $v_1 = v_1(B_{x^*})$ is the reluctivity, frequently nonlinear, in the x^* direction, and

$$H_{y^*} = v_2 B_{y^*}$$

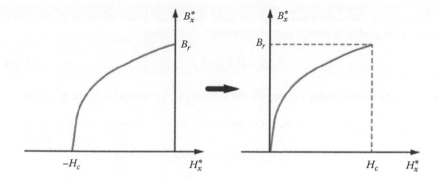

FIGURE 4.25 Treatment of permanent magnets.

Here, $\nu_2 = \nu_{recoil}$ is the reluctivity on the normal direction of magnetization of the permanent magnet. This magnetization curve is frequently linear and closed, which forms the magnetic permeability of the air (or vacuum).

This constitutive relation can be represented in matrix form as follows:

$$\begin{bmatrix} H_{x^*} \\ H_{y^*} \end{bmatrix} = \begin{pmatrix} \nu_1 & 0 \\ 0 & \nu_2 \end{pmatrix} \begin{bmatrix} B_{x^*} \\ B_{y^*} \end{bmatrix} \tag{4.54}$$

or in the compact form

$$\left[H^* \right] = \left[\nu \right] \left[B^* \right]$$

In the mathematical treatment by FEM, we adopt a global coordinates system (x, y), which necessarily does not match the system (x^*, y^*). Due to this, we have to apply a transformation on the coordinates system in order to match the reference systems.

Figure 4.26 shows both reference systems. α being the angle between the x^*-axis of the local referential of the magnet and the x-axis from the global reference system.

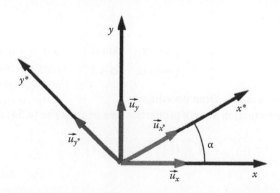

FIGURE 4.26 Coordinates systems.

$\vec{B}$ being the bidimensional magnetic field in a given point of the domain under study, on the global system it can be expressed as follows:

$$\vec{B} = B_x \vec{u}_x + B_y \vec{u}_y \tag{4.55}$$

The relations between unit vectors on the systems (x^*, y^*) and (x, y) are given by

$$\vec{u}_x = \cos\alpha\vec{u}_{x^*} - \sin\alpha\vec{u}_{y^*}$$

$$\vec{u}_y = \sin\alpha\vec{u}_{x^*} - \cos\alpha\vec{u}_{y^*}$$

Replacing $\vec{u}_x$ and $\vec{u}_y$ by their values in (4.55), we get

$$\vec{B} = B_x(\cos\alpha\vec{u}_{x^*} - \sin\alpha\vec{u}_{y^*}) + B_y(\sin\alpha\vec{u}_{x^*} - \cos\alpha\vec{u}_{y^*})$$

or

$$\vec{B} = B_x^* \vec{u}_{x^*} + B_y^* \vec{u}_{y^*}$$

where

$$B_x^* = B_x \cos\alpha + B_y\sin\alpha \quad \text{and} \quad B_y^* = -B_x\sin\alpha + B_y\cos\alpha$$

This is written in the matrix form as follows:

$$\begin{bmatrix} B_{x^*} \\ B_{y^*} \end{bmatrix} = \begin{pmatrix} \cos\alpha & \sin\alpha \\ -\sin\alpha & \cos\alpha \end{pmatrix} \begin{bmatrix} B_x \\ B_y \end{bmatrix}$$

or in the compact form

$$\begin{bmatrix} B^* \end{bmatrix} = \begin{bmatrix} T \end{bmatrix}\begin{bmatrix} B \end{bmatrix}$$

where, the matrix

$$\begin{bmatrix} T \end{bmatrix} = \begin{pmatrix} \cos\alpha & \sin\alpha \\ -\sin\alpha & \cos\alpha \end{pmatrix}$$

is the transformation matrix from coordinates.

Using the transformation matrix properties, we can rewrite (4.54) as follows:

$$\begin{bmatrix} T \end{bmatrix}\begin{bmatrix} H \end{bmatrix} = \begin{bmatrix} v \end{bmatrix}\begin{bmatrix} T \end{bmatrix}\begin{bmatrix} B \end{bmatrix} \tag{4.56}$$

or

$$\begin{bmatrix} H \end{bmatrix} = \begin{bmatrix} T \end{bmatrix}^{-1}\begin{bmatrix} v \end{bmatrix}\begin{bmatrix} T \end{bmatrix}\begin{bmatrix} B \end{bmatrix} \tag{4.57}$$

Developing this result, we get

$$\begin{bmatrix} H_x \\ H_y \end{bmatrix} = \begin{pmatrix} \nu_{xx} & \nu_{xy} \\ \nu_{xy} & \nu_{yy} \end{pmatrix} \begin{bmatrix} B_x \\ B_y \end{bmatrix} \tag{4.58}$$

where

$$\nu_{xx} = \nu_1 \cos^2 \alpha + \nu_2 \sin^2 \alpha$$

$$\nu_{yy} = \nu_2 \cos^2 \alpha + \nu_1 \sin^2 \alpha$$

$$\nu_{xy} = (\nu_1 - \nu_2) \cos \alpha \sin \alpha$$

Applying this transformation to the coercive magnetic intensity field (H_c), which only contains the (x^*) component, we can write

$$\begin{bmatrix} H_c \\ 0 \end{bmatrix} = [T] \begin{bmatrix} H_{cx} \\ H_{cy} \end{bmatrix} \tag{4.59}$$

which in turn results in

$$H_{cx} = H_c \cos \alpha \quad \text{and} \quad H_{cy} = H_c \sin \alpha$$

Reassuming the calculation of E_1^e expressed in (4.31) for magnetostatics, we get

$$E_1^e = \int_{POS} \vec{H} \cdot d\vec{l} = H_x \Delta x - H_y \Delta y$$

and from (4.58), we can write

$$H_x = \nu_{xx} B_x + \nu_{xy} B_y$$

$$H_y = \nu_{xy} B_x + \nu_{yy} B_y$$

It then results in

$$E_1^e = \int_{POS} \vec{H} \cdot d\vec{l} = (\nu_{xx} B_x + \nu_{xy} B_y) \Delta x - (\nu_{xy} B_x + \nu_{yy} B_y) \Delta y$$

Remembering that

$$B_x = \frac{1}{2\Delta} (c_1 A_1 + c_2 A_2 + c_3 A_3)$$

$$B_y = -\frac{1}{2\Delta} (b_1 A_1 + b_2 A_2 + b_3 A_3)$$

$$\Delta x = c_1 / 2 \quad \text{and} \quad \Delta y = b_1 / 2$$

we finally obtain

$$E_1^e = \frac{1}{4\Delta}\{[v_{xx}c_1c_1 + v_{yy}b_1b_1 - v_{xy}(b_1c_1 + c_1b_1)]A_1$$

$$+[v_{xx}c_1c_2 + v_{yy}b_1b_2 - v_{xy}(b_1c_2 + c_1b_2)]A_2 \qquad (4.60)$$

$$+[v_{xx}c_1c_3 + v_{yy}b_1b_3 - v_{xy}(b_1c_3 + c_1b_3)]A_3\}$$

Following the same procedure, we obtain the expressions for E_2^e and E_3^e, which will result in expressions similar to those in (4.60).

Completing the partial circulations and expressing the result in matrix form, we can write

$$\begin{bmatrix} E_1^e \\ E_2^e \\ E_3^e \end{bmatrix} = \begin{pmatrix} S_{11} & S_{12} & S_{13} \\ S_{21} & S_{22} & S_{23} \\ S_{31} & S_{32} & S_{33} \end{pmatrix} \cdot \begin{bmatrix} A_1 \\ A_2 \\ A_3 \end{bmatrix} \qquad (4.61)$$

where

$$S_{ij} \frac{1}{4\Delta}[v_{xx}c_ic_j + v_{yy}b_ib_j - v_{xy}(b_ic_j + c_ib_j)] \quad i, j = 1, 2, 3 \qquad (4.62)$$

Note that in the matrix from, the element maintains its symmetry. Besides, the matrix of the element for the isotropic case is obtained from (4.62), which gives

$$v_{xx} = v_{yy} = v \text{ and } v_{xy} = 0$$

In the case of elements whose magnetic material is a permanent magnet, the actions vector is obtained based on the coercive magnetic intensity field H_c.

Figure 4.27 shows the generic finite element used for the establishment of the general term of the element's matrix.

R_1^e being the circulation section of H_c on the path POS with H_c constant inside the element, we get

$$R_1^e = H_{cx}\Delta x - H_{cy}\Delta y$$

or

$$R_1^e = \frac{1}{2}(H_{cx}c_1 - H_{cy}b_1) \qquad (4.63)$$

For the remaining sections, we get

$$R_2^e = \frac{1}{2}(H_{cx}c_2 - H_{cy}b_2) \qquad (4.64)$$

$$R_3^e = \frac{1}{2}(H_{cx}c_3 - H_{cy}b_3) \qquad (4.65)$$

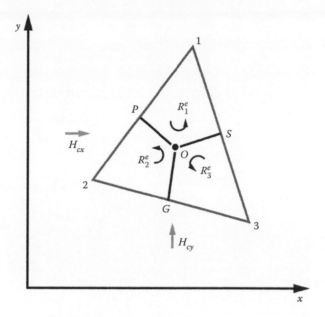

FIGURE 4.27   Actions vector for permanent magnets.

Therefore, the vector of the actions for elements of permanent magnets will be given by

$$
\begin{bmatrix} R_1^e \\ R_2^e \\ R_3^e \end{bmatrix} = \frac{1}{2} \begin{bmatrix} H_{cx}c_1 - H_{cy}b_1 \\ H_{cx}c_2 - H_{cy}b_2 \\ H_{cx}c_3 - H_{cy}b_3 \end{bmatrix}
$$

The treatment for assembling the global actions vector is performed using the same procedure.

5 Finite Element Method for Time-Dependent Electromagnetic Fields

5.1 INTRODUCTION

Studies involving electromagnetic phenomena depending on time are classified into two classes. The first refers to the study of quasi-static electromagnetic fields, where the time variation rate of electromagnetic field is sufficiently low such that the effects of displacement current can be neglected. The second refers to the study of time-dependent electromagnetic field, where the displacement current cannot be neglected. The problems involved in the first class are related to the industrial electromagnetic problems, for example, those present on electric machines and transients in operation of short transmission lines. The association between electromagnetism and theory of electric circuits also belongs to this class of studies. The second class has a particular interest in studies involving the propagation of electromagnetic fields, which is a characteristic of electromagnetic waves. Due to the large industrial progress toward the end of the last century, different techniques involving time-dependent electromagnetic fields studies were incorporated to study industrial problems, like those issued from the digitalization of the control of both rotating and static electrical machines. This is the case for both electromagnetic interference (EMI) and electromagnetic compatibility (EMC) due to the intensive use of not only of high-frequency equipment but also digital instruments sensible to the actions of time-dependent fields [15].

5.2 QUASI-STATIC ELECTROMAGNETIC FIELD

The quasi-static electromagnetic field, modeled by FEM, transforms Maxwell's second equation in the integral form to a time-dependent algebraic equations system. For solving it, the problem requires not only a time-integration procedure but also the introduction of boundary conditions to the nodes at each step. In addition, the initial conditions must be incorporated for describing the state of the system before the occurrence of the disturbances which are object of studies [38,70].

Before the application of the methodology, a small manipulation is done on Maxwell's equations for adapting them to our methodology. In the present case, as mentioned earlier, the displacement current is neglected, so that Maxwell's second equation in the integral form becomes

$$\oint_C \vec{H} \cdot d\vec{l} = \int_S \vec{J} \cdot d\vec{S}.$$ (5.1)

Recalling that

$$\vec{J} = \sigma\vec{E} \quad \text{and} \quad \vec{E} = -\nabla V - \frac{\partial\vec{A}}{\partial t},$$

Equation 5.1 is rewritten as

$$\oint_C \vec{H} \cdot d\vec{l} = \int_S \sigma\left(-\nabla V - \frac{\partial\vec{A}}{\partial t}\right) \cdot d\vec{S}.$$

The current density vector $\vec{J}_f$ is defined as

$$\vec{J}_f = -\sigma\nabla V, \tag{5.2}$$

which is called *source current density vector*. Thus, we arrive at the following expression from Maxwell's second equation in the integral form for the quasi-static electromagnetic phenomena

$$\oint_C \vec{H} \cdot d\vec{l} = \int_S \vec{J}_f - \int_S \sigma\frac{\partial\vec{A}}{\partial t} \cdot d\vec{S}. \tag{5.3}$$

The treatment of this equation by our methodology is similar, on the whole, to the development elaborated on magnetostatics studies. Applying our methodology to the generic finite element, we get the following matrix expressions:

1. The right-hand-side term generates the following circulation sections around the nodes

$$\begin{bmatrix} E_1^e \\ E_2^e \\ E_3^e \end{bmatrix} = \frac{\nu}{4\Delta}\begin{pmatrix} b_1 b_1 + c_1 c_1 & b_1 b_2 + c_1 c_2 & b_1 b_3 + c_1 c_3 \\ b_2 b_1 + c_2 c_1 & b_2 b_2 + c_2 c_2 & b_2 b_3 + c_2 c_3 \\ b_3 b_1 + c_3 c_1 & b_3 b_2 + c_3 c_2 & b_3 b_3 + c_3 c_3 \end{pmatrix}\begin{bmatrix} A_1 \\ A_2 \\ A_3 \end{bmatrix}. \tag{5.4}$$

2. The first term from the right-hand side will produce the following action vectors

$$\begin{bmatrix} I_1^e \\ I_2^e \\ I_3^e \end{bmatrix} = \begin{bmatrix} J_f\Delta/3 \\ J_f\Delta/3 \\ J_f\Delta/3 \end{bmatrix}. \tag{5.5}$$

Next we establish the matrix equation that is associated with the term of the right-hand side of the Equation 5.3. In order to achieve that, let us consider the generic finite element shown in Figure 5.1.

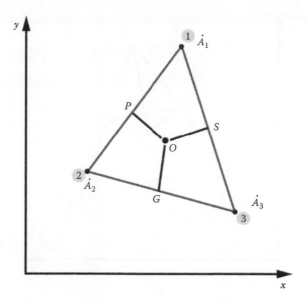

FIGURE 5.1   Generic finite element.

The aim is to evaluate the flux of $\partial \vec{A}/\partial t$ over the polygonal surface that involves each element's nodes, which passes through the centroid and through the middle of sides of the element.

Each finite element contributes to a portion of the flux on three different surfaces involving nodes 1, 2, and 3. Considering the nodes corresponding to the surface, which involves node 1, it is necessary to evaluate the flux of $\partial \vec{A}/\partial t$ over the surface surrounded by the sequence of points *1POS1*. As the component (z) of the magnetic potential vector is linearly variable inside the element, it is possible to evaluate the average value of this vector over the surface *1POS1*. As the magnetic potential vector is linear inside the element, the time variation $\dot{A} = \partial A/\partial t$ shows similar behavior so that the values of these variations on vertexes of the polygon *1POS1* are as follows

Node 1: $\dot{A}_1 = \partial A_1/\partial t$
Node P (the middle point of nodes 1 and 2): $\dot{A}_P = 1/2(\partial A_1/\partial t + \partial A_2/\partial t) = 1/2(\dot{A}_1 + \dot{A}_2)$
Node O (the centroid): $\dot{A}_o = 1/3(\dot{A}_1 + \dot{A}_2 + \dot{A}_3)$
Node S (the middle point of nodes 1 and 3): $\dot{A}_S = 1/2(\dot{A}_1 + \dot{A}_3)$.

We assume a representative value as an average value for the time variation of the magnetic potential vector inside the polygon, which is evaluated in the middle of the segment *PS*—point K, and this value is given by (Figure 5.2)

$$\dot{A}_K = \frac{1}{2}\left(A_P + A_S \right) = \frac{1}{2}\left(\frac{\dot{A}_1 + \dot{A}_2}{2} + \frac{\dot{A}_1 + \dot{A}_3}{2} \right) = \frac{1}{2}\left(\frac{2\dot{A}_1 + \dot{A}_2 + \dot{A}_3}{2} \right). \qquad (5.6)$$

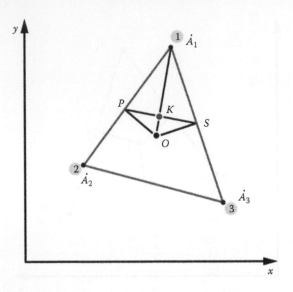

FIGURE 5.2   Average value of $\dot{A}$.

Thus, the $\dot{A}$ flux over the surface delimited by the polygon *1POS1* is given by

$$F_1^e = \dot{A}_K \frac{\Delta}{3} = \frac{\Delta}{12}(2\dot{A}_1 + \dot{A}_2 + \dot{A}_3).$$

Following the same procedure for the other surfaces, we get

$$F_2^e = \frac{\Delta}{12}(\dot{A}_1 + 2\dot{A}_2 + \dot{A}_3)$$

$$F_3^e = \frac{\Delta}{12}(\dot{A}_1 + \dot{A}_2 + 2\dot{A}_3).$$

Representing these results in a matrix form, we get

$$\begin{bmatrix} F_1^e \\ F_2^e \\ F_3^e \end{bmatrix} = \frac{\Delta}{12} \begin{pmatrix} 2 & 1 & 1 \\ 1 & 2 & 1 \\ 1 & 1 & 2 \end{pmatrix} \begin{bmatrix} \dot{A}_1 \\ \dot{A}_2 \\ \dot{A}_3 \end{bmatrix}. \tag{5.7}$$

Thus, using matrix expressions (5.4), (5.5), and (5.7) we can represent Maxwell's second equation applied to the polygon involving the node (*i*) as

$$\sum_{e=1}^{NE} E_i^e = \sum_{e=1}^{NE} I_i^e - \sum_{e=1}^{NE} \sigma F_i^e \quad i = 1, ..., NN$$

or as

$$\sum_{e=1}^{NE} \sigma F_i^e + \sum_{e=1}^{NE} E_i^e = \sum_{e=1}^{NE} I_i^e \quad i=1,\ldots,NN \tag{5.8}$$

This could be globally represented in a matrix form

$$\left[C\right]\left[\dot{A}\right]+\left[S\right]\left[A\right]=\left[I\right] \tag{5.9}$$

for which elements of the generic matrixes are given by

$$C_{ij} = \begin{cases} \dfrac{\sigma\Delta}{6} & \text{for } i=j \\[2mm] \dfrac{\sigma\Delta}{12} & \text{for } i\neq j \end{cases}, \tag{5.10}$$

$$S_{ij} = \frac{\nu}{4\Delta}(b_i b_j + c_i c_j), \tag{5.11}$$

$$I_j = J_f \frac{\Delta}{3}. \tag{5.12}$$

To solve Equation 5.9 a numerical time-domain integration scheme is required so that the initial conditions, that is, magnetic potential values at $t = t_0$, could be known.

The methodology, which we discussed, is used to evaluate the transient behavior of electromechanical devices, for example, which occur in electric machines and other electromechanical devices. This methodology is very suitable for modeling simultaneously electric motor with its respective static convert drive, because the association of the magnetic study associated with the simulation techniques of electric circuits allow us the evaluation of the behavior, not only the magnetic performance of the device but also the performance of the static converter that drive it. To know more about advanced techniques of electric circuit's coupling with the magnetic circuit, the reader is suggested to refer Cardoso, Bastos, Abe [18,38,65].

As an example, we chose to simulate a single planar electromagnetic actuator, which is suitable for showing the power of this formalism in the transient electromagnetic phenomenon.

The method involves the application of a step voltage in the actuator coil having a solid conductive armature. Figure 5.3 shows the armature eddy current effect in equipotential lines soon after the coil connection to the source is established. Not only it is possible to evaluate equipotential lines but also all quantities associated with the phenomenon at each step, such as force, inductance variation, loss, etc.

Figure 5.4 shows the same result at $t = 0.75$ s when the steady state is reached. Note that the skin effect due to the eddy current does not exist anymore, and no eddy current effect is detectable.

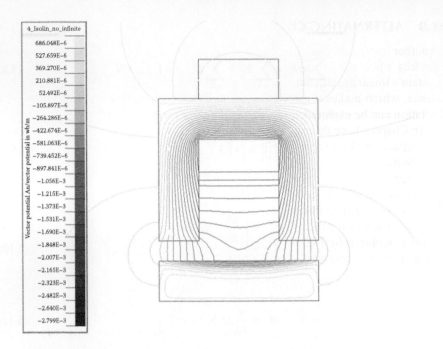

FIGURE 5.3 Eddy current effect due to step voltage application in the coil, $t = 0_+$.

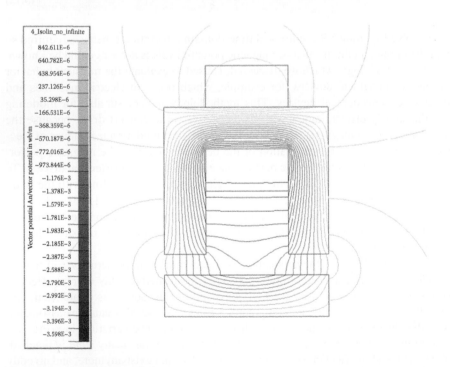

FIGURE 5.4 Eddy current effect due to step voltage application in the coil, $t = 0.75$ s.

5.3 ALTERNATING CURRENT STEADY-STATE MAGNETODYNAMICS

Another important quasi-static time-dependent analysis is the behavior of electric devices when fed by an alternating current source. In this case, assuming the system's linearity, all voltages, currents, and fields present sinusoidal time dependence, which makes it easier to solve the problem since advantages of complex notation can be explored.

In Chapter 3, we discuss the complex representation of sinusoidal quantities and its impact on the steady-state solution of sinusoidal time-dependent differential equations. Note that a sinusoidal time-dependent differential equation solved using a complex notation is transformed to an algebraic equation having complex coefficients called phasors [50,55].

Thus, the matrix differential equation (5.9) can be rewritten applying the complex notation transform called the "frequency domain." For this, every time variation $(\partial/\partial t)$ is replaced by $(j\omega)$ on the frequency domain. Thus, the matrix differential equation (5.9) is rewritten as

$$j\omega[C]\left[\hat{A}\right]+[S]\left[\hat{A}\right]=\left[\hat{I}\right]$$

or as

$$\left(j\omega[C]+[S]\right).\left[\hat{A}\right]=\left[\hat{I}\right], \tag{5.13}$$

where $\left[\hat{A}\right]$ and $\left[\hat{I}\right]$ are column vectors that correspond to the magnitude and phase of the corresponding quantity (its phasor), respectively. Thus, the solution to the problem lies in solving the following complex equation system

$$\left[\hat{S}\right].\left[\hat{A}\right]=\left[\hat{I}\right], \tag{5.14}$$

where

$$\left[\hat{S}\right]=[S]+j\omega[C]. \tag{5.15}$$

5.3.1 NUMERICAL EXAMPLE: MAGNETODYNAMICS

It is about a problem involved while assessing the eddy current density on bars of a squirrel cage of an induction motor. This problem was originally highlighted by Silvester in [4,5] and solved by a step by step methodology, detailed in [11].

The next figure shows the cross section of a slot discretized into a first-order triangular finite element (Figure 5.5). Due to its symmetry, only half of the domain is studied. Table 5.1 presents nodes of each element and the corresponding current density imposed by the source that is normal to the plane of the figure. Table 5.2 provides nodal coordinates in centimeters. A ferromagnetic material is considered ideal and the conductor inside the slot is copper, which has a permeability value equal to air's ($4\pi \times 10^{-7}$ H/m) and a conductivity value of 5.8×10^7 (S/m). To find a solution, as similar as to the real case, the frequency of rotor currents is assumed to

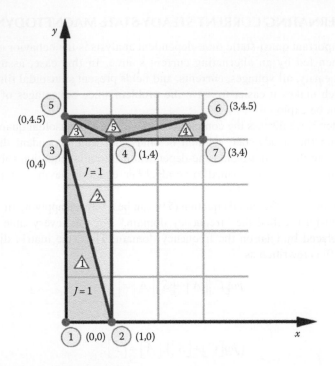

FIGURE 5.5   Slot of an induction motor.

TABLE 5.1

Node Incidence from Elements and Impressed Current Density

Element	Node 1	Node 2	Node 3	J_i (A/mm²)
1	1	2	3	1
2	2	4	3	1
3	3	4	5	0
4	4	7	6	0
5	4	6	5	0

TABLE 5.2

Node Coordinates and Closed Path Conditions

Node	X	Y	A (Wb/m)
1	0	0	—
2	1	0	—
3	0	4	—
4	1	4	—
5	0	4.5	—
6	3	4.5	0
7	3	4	0

be a slip frequency of 3 Hz. At this frequency the penetration depth is about 4 cm, comparable to the slot dimensions, which makes it possible to devise a solution using a coarse finite element mesh.

As a boundary condition, the component (z) of the magnetic potential vector is null on points 6 and 7. Therefore, $A_6 = A_7 = 0$.

5.3.2 Calculus of the Matrix Element

The matrix $[S^1]$ from element (1) calculated from Equation 5.11is given by (Table 5.3)

$$[S^1] = \frac{795,774.7}{4 \times 0.0002} \begin{pmatrix} 0.0017 & -0.0016 & -0.0001 \\ -0.0016 & 0.0016 & 0 \\ -0.0001 & 0 & 0.0001 \end{pmatrix}.$$

This same procedure is repeated for all other elements that make it possible obtain all other matrixes $[S^e]$ from elements 2, 3, 4, and 5.

Since the problem is magnetodynamics with an alternating current, we also need the matrix $[C^e]$ whose elements are evaluated from Equation 5.10 resulting

$$[C^1] = \frac{5.8 \times 10^7 \times 0.002}{12} \begin{pmatrix} 2 & 1 & 1 \\ 1 & 2 & 1 \\ 1 & 1 & 2 \end{pmatrix}.$$

Next we evaluate the complement matrix of the element (1), $[\hat{S}^1]$, which is given by

$$[\hat{S}^1] = [S^1] + j\omega[C^1].$$

Note that $\omega = 2\pi f = 2\pi 3 = 6\pi$, which results in

$$[\hat{S}^1] = 0.994 \times 10^9 \begin{pmatrix} 0.0017 & -0.0016 & -0.0001 \\ -0.0016 & 0.0016 & 0 \\ -0.0001 & 0 & 0.0001 \end{pmatrix} + j18,221.24 \begin{pmatrix} 2 & 1 & 1 \\ 1 & 2 & 1 \\ 1 & 1 & 2 \end{pmatrix}.$$

This procedure shall be repeated for all the other mesh's elements.

TABLE 5.3

Element Data (1)

Local numbering	1	2	3
Global numbering	1	2	3
X (m)	0	0.01	0
Y (m)	0	0	0.04
b's (m)	−0.04	0.04	0
c's (m)	−0.01	0	0.01

Note: Element's area: $\Delta = (b_1 c_2 - b_2 c_1)/2 = 0.2 \times 10^{-3} \ m^2$.

5.3.3 EVALUATION OF ACTION VECTORS

As there are only sources on elements (1) and (2), only vectors $[\hat{I}^1]$ and $[\hat{I}^2]$ are calculated. Imposing null vector, the phase of source current density that results is as follows

$$\left[\hat{I}^1\right] = \left[\hat{I}^2\right] = \frac{1 \times 10^{-3} \times 0.0002}{3} \begin{bmatrix} 1 \\ 1 \\ 1 \end{bmatrix} = \begin{bmatrix} 66.667 \\ 66.667 \\ 66.667 \end{bmatrix}.$$

5.3.4 ASSEMBLY OF GLOBAL MATRIX AND INTRODUCTION OF BOUNDARY CONDITIONS

Applying the algorithm of assembling the global matrix explained in Chapter 4 and also imposing boundary conditions and eliminating the nodes, the magnetic potentials of which are known, we get the following complex equation system

$$10^6 \begin{bmatrix} 1.691+j0.0036 & -1.592+j0.018 & -0.099+j0.0018 & 0 & 0 & 0 & 0 \\ -1.592+j0.018 & 1.691+j0.0073 & j0.0036 & -0.099+j0.0180 & 0 & 0 & 0 \\ -0.099+j0.0018 & j0.0036 & 2.686+j0.073 & -1.791+j0.018 & -0.796 & 0 & 0 \\ 0 & -0.099+j0.0180 & -1.791+j0.018 & 4.377+j0.036 & -1.592 & 0 & 0 \\ 0 & 0 & -0.796 & -1.592 & 1.923 & 0 & 0 \\ 0 & 0 & 0 & 0 & 0 & 1 & 0 \\ 0 & 0 & 0 & 0 & 0 & 0 & 1 \end{bmatrix}$$

$$\times \begin{bmatrix} \hat{A}_1 \\ \hat{A}_1 \\ \hat{A}_1 \\ \hat{A}_1 \\ \hat{A}_1 \\ \hat{A}_1 \\ \hat{A}_1 \end{bmatrix} = \begin{bmatrix} 66.667 \\ 133.33 \\ 133.33 \\ 66.667 \\ 0 \\ 0 \\ 0 \end{bmatrix}.$$

The solution provides us values of the component (z) of the magnetic potential vector (Wb/m), which are given as

$$\hat{A}_1 = 0.386 - j0.931;$$

$$\hat{A}_2 = 0.389 - j0.936;$$

$$\hat{A}_3 = 0.293 - j0.585;$$

$$\hat{A}_4 = 0.254 - j0.506;$$

$$\hat{A}_5 = 0.332 - j0.661;$$

$$\hat{A}_6 = 0$$

$$\hat{A}_7 = 0.$$

5.3.5 EVALUATING THE EDDY CURRENT INDUCED ON ELEMENT (1)

The eddy current on elements where the current is not imposed by a source is initiated through the assessment of eddy current density expressed by

$$\hat{J}_{ind} = -j\omega\sigma\hat{A}.$$

Note the eddy current density is linearly dependent inside the element. An average value for this quantity can be obtained by calculating the eddy current density on the element's centroid. To achieve this, it is initially assessed that the average value of the magnetic potential vector is given by

$$\hat{A}_{med}^1 = \frac{\hat{A}_1 + \hat{A}_2 + \hat{A}_3}{3},$$

which results in

$$\hat{A}_{med}^1 = 0.3562 + j0.8174 \quad (\text{Wb/m})$$

$$A_{med}^1 = 0.89164 \quad (\text{Wb/m}),$$

so that

$$J_{ind} = 6\pi \times 5.8 \times 10^7 \times 0.89164 \times 10^{-6} = 0.975 \quad (\text{A/mm}^2).$$

6 Finite Element Method for Axisymmetric Geometries

6.1 INTRODUCTION

Several electromagnetic devices present cylindrical geometry; this type of symmetry is also called axisymmetric geometry, an anachronism of axial geometry. These devices present, necessarily, a rotation axis so that any plane passing by this axis presents the same cross section. Some examples of devices that present axisymmetric geometry are cylindrical actuators, transmission line isolators, and coaxial cables.

In axisymmetric geometry, the domain under study is a cross section that contains necessarily the rotation axis that agrees with the axis (z) of the cylindrical coordinates system adopted in the study.

As the plane is symmetrical, both the electric and magnetic fields have only two components at the cross section. Thus, in axisymmetric geometry, the electromagnetic fields present only radial (*axis r*) and axial (*axis z*) components [30].

6.2 ELECTROSTATICS IN AXISYMMETRIC GEOMETRY

In axisymmetric electrostatics, all conductors present cylindrical geometry. A typical axisymmetric electrostatics problem consists of a set of energized "toroidal" conductors separated by ideal dielectrics, as shown in Figure 6.1.

The starting point for this development is Maxwell's fourth equation, the electrostatics Gauss' law, given by

$$\oint_{\Sigma} \vec{D} \cdot d\vec{S} = Q_i \tag{6.1}$$

The surface Σ in Equation 6.1 in axisymmetric studies is a toroid with a cross section composed of the same polygonal line, that is, the polygonal line that involves each mesh's node linking the centroid to the middle of the edges.

Figure 6.2 shows a generic element extracted from the finite elements mesh. In fact, this figure is a triangular cross section of a toroidal finite element, the revolution axis of which is the (z)-axis of the cylindrical coordinates system.

Maxwell's fourth equation (6.1), applied to the toroidal closed surface of polygonal cross section, will be evaluated in sections, each one of these sections is contained on the elements that constitute it. Thus, lines *PO* and *OS* in Figure 6.2 is a section of the closed surface involving node (1); *PO* and *OG* is a section of the closed surface

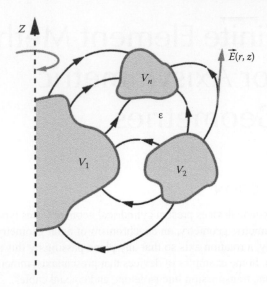

FIGURE 6.1 Typical problem of axisymmetric electrostatics.

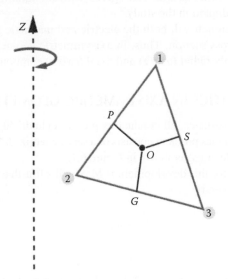

FIGURE 6.2 Axisymmetric generic finite element.

involving node (2); and finally, *GO* and *OS* is a section of the same closed surface involving node (3). This geometry is the cross section of a rotation volume over the axis (*z*) of the coordinates system.

Thus, the integral

$$E_1^e = \int\limits_{POS} \vec{D} \cdot d\vec{S} \qquad (6.2)$$

is a portion of the integral

$$\oint_\Sigma \vec{D} \cdot d\vec{S} = Q_1$$

which is Maxwell's fourth equation applied to the closed surface with those characteristics involving node (1) of the generic finite element (e), denoted by Σ_1.

The electric field inside the element in *cylindrical coordinates* is given by

$$\vec{E} = -\nabla V = -\frac{\partial V}{\partial r}\vec{u}_r - \frac{\partial V}{\partial z}\vec{u}_z \qquad (6.3)$$

Expressing the distribution of $V(r,z)$ inside the element with a linear interpolation of its values on vertexes in the same way we made before, we can write

$$V(r,z) = N_1 V_1 + N_2 V_2 + N_3 V_3 \qquad (6.4)$$

where

$$N_i = \frac{1}{2\Delta}(a_i + b_i r + c_i z) \quad i = 1,2,3 \qquad (6.5)$$

with

$$a_1 = r_2 z_3 - r_3 z_2; \quad b_1 = z_2 - z_3; \quad c_1 = r_3 - r_2; \quad \Delta = (b_1 c_2 - b_2 c_1)/2$$

The others coefficients are obtained by cyclical rotation of the indexes, where Δ is the finite element's area.

Thus, we can write

$$E_r = -\frac{\partial V}{\partial r} = -\frac{1}{2\Delta}(b_1 V_1 + b_2 V_2 + b_3 V_3)$$

$$E_z = -\frac{\partial V}{\partial z} = -\frac{1}{2\Delta}(c_1 V_1 + c_2 V_2 + c_3 V_3) \qquad (6.6)$$

as $\vec{D} = \varepsilon \vec{E}$ results in

$$D_r = \varepsilon E_r = -\frac{\varepsilon}{2\Delta}(b_1 V_1 + b_2 V_2 + b_3 V_3)$$

$$D_z = \varepsilon E_z = -\frac{\varepsilon}{2\Delta}(c_1 V_1 + c_2 V_2 + c_3 V_3) \qquad (6.7)$$

These expressions show that the displacement vector $\vec{D}$ is constant inside the element, so that the integral (6.2) is reduced to

$$E_1^e = \int_{POS} \vec{D} \cdot d\vec{S} = \vec{D} \cdot \Delta\vec{S} \qquad (6.8)$$

where

$$\vec{D} = D_r\vec{u}_r + D_z\vec{u}_z$$

and

$$\Delta\vec{S} = 2\pi r_b \Delta z(-\vec{u}_r) + 2\pi r_b \Delta r(-\vec{u}_z)$$

Note that the elementary area's vector is always normal and leaving the surface. In this case, this vector shall be normal to the polygonal surface that links the points P and S, as shown in Figure 6.3.

In addition, in this case we have the following relations:

$$\Delta z = z_p - z_S = \frac{1}{2}(z_2 - z_3) = \frac{b_1}{2}; \ \Delta r = r_S - r_p = \frac{1}{2}(r_3 - r_2) = \frac{c_1}{2}$$

The following expression is the radius of the centroid of the generic finite element:

$$r_b = (r_1 + r_2 + r_3)/3$$

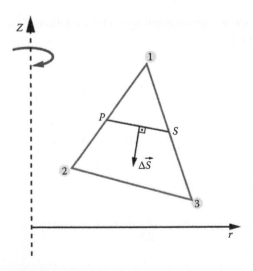

FIGURE 6.3 Elementary area's vector.

Replacing these values in Equation 6.8 gives

$$E_1^e = \frac{2\pi r_b \varepsilon}{4\Delta}[(b_1 b_1 + c_1 c_1)V_1 + (b_1 b_2 + c_1 c_2)V_2 + (b_1 b_3 + c_1 c_3)V_3] \qquad (6.9)$$

Following the same procedure for the portions of the surfaces that involve nodes (2) and (3), we obtain

$$E_2^e = \frac{2\pi r_b \varepsilon}{4\Delta}[(b_2 b_1 + c_2 c_1)V_1 + (b_2 b_2 + c_2 c_2)V_2 + (b_2 b_3 + c_2 c_3)V_3] \qquad (6.10)$$

$$E_3^e = \frac{2\pi r_b \varepsilon}{4\Delta}[(b_3 b_1 + c_3 c_1)V_1 + (b_3 b_2 + c_3 c_2)V_2 + (b_3 b_3 + c_3 c_3)V_3] \qquad (6.11)$$

The results in Equations 6.9 through 6.11 can be written in matrix form as follows:

$$\begin{bmatrix} E_1^e \\ E_2^e \\ E_3^e \end{bmatrix} = \frac{2\pi r_b \varepsilon}{4\Delta} \begin{pmatrix} b_1 b_1 + c_1 c_1 & b_1 b_2 + c_1 c_2 & b_1 b_3 + c_1 c_3 \\ b_2 b_1 + c_2 c_1 & b_2 b_2 + c_2 c_2 & b_2 b_3 + c_2 c_3 \\ b_3 b_1 + c_3 c_1 & b_3 b_2 + c_3 c_2 & b_3 b_3 + c_3 c_3 \end{pmatrix} \cdot \begin{bmatrix} V_1 \\ V_2 \\ V_3 \end{bmatrix} \qquad (6.12)$$

The matrix

$$[C^e] = \frac{2\pi r_b \varepsilon}{4\Delta} \begin{pmatrix} b_1 b_1 + c_1 c_1 & b_1 b_2 + c_1 c_2 & b_1 b_3 + c_1 c_3 \\ b_2 b_1 + c_2 c_1 & b_2 b_2 + c_2 c_2 & b_2 b_3 + c_2 c_3 \\ b_3 b_1 + c_3 c_1 & b_3 b_2 + c_3 c_2 & b_3 b_3 + c_3 c_3 \end{pmatrix} \qquad (6.13)$$

is called element's matrix (e). Its characteristics, as already highlighted in the previous chapters, are both singular and symmetric and its dimension is *Faraday* (F).

6.2.1 ELECTRIC CHARGE INTERNAL TO SURFACE

The right-hand side of (6.1) is the internal charge of the surface Σ. This amount of electric charge is distributed over all the elements that have the highlighted node as vertex. Therefore, for the Σ involving node (1), called Σ_1, the portion of the electric charges inside this surface that belong to the element (e) is 1/3 the total amount of electric charge in this element. Thus, we can write

$$Q_1^e = \frac{1}{3}Q_i$$

Noting that

$$Q_i = \rho \text{Volume} = \rho 2\pi r_b \Delta$$

gives

$$Q_1^e = \frac{\rho 2\pi r_b \Delta}{3} \tag{6.14}$$

The same values in (6.14) are obtained for the portions of the total amount of the internal charges to surfaces involving nodes (2) and (3), and we can represent these portions in matrix form as follows:

$$\begin{bmatrix} Q_1^e \\ Q_2^e \\ Q_3^e \end{bmatrix} = \begin{bmatrix} \rho 2\pi r_b \Delta/3 \\ \rho 2\pi r_b \Delta/3 \\ \rho 2\pi r_b \Delta/3 \end{bmatrix} \tag{6.15}$$

Finally, the application of Maxwell's fourth equation on a closed surface involving the generic node (i) of the mesh can be written as follows:

$$\sum_{e=1}^{NE} E_i^e = \sum_{e=1}^{NE} Q_i^e \quad i = 1, 2, \ldots, NN \tag{6.16}$$

where
NE is the total number of elements from mesh
NN is the total number of nodes

It should be noted again that the terms from sums indicated in Equation 6.16 will only have non-null values on the elements (e's) that have node (i) as a vertex. This expression also generates a system of NN equations with NN unknowns, where the unknowns are the electric potential of the element's nodes. This can be written as follows:

$$[C][V] = [Q] \tag{6.17}$$

The matrix [C] presents the same characteristics of the global matrices discussed in the previous chapters, that is, it is symmetric, sparse, and singular.

6.3   ELECTROKINETICS IN AXISYMMETRIC GEOMETRY

As discussed in Chapter 4, Table 6.1 shows the duality between electrokinetics and electrostatics.

Therefore, for obtaining the elements' matrix in electrokinetics studies, it is sufficient to replace the variables and the parameters as shown in Table 6.1. Applying these convenient substitutions, the axisymmetric element matrix for electrokinetics is

$$\begin{bmatrix} E_1^e \\ E_2^e \\ E_3^e \end{bmatrix} = \frac{2\pi r_b \sigma}{4\Delta} \begin{pmatrix} b_1b_1 + c_1c_1 & b_1b_2 + c_1c_2 & b_1b_3 + c_1c_3 \\ b_2b_1 + c_2c_1 & b_2b_2 + c_2c_2 & b_2b_3 + c_2c_3 \\ b_3b_1 + c_3c_1 & b_3b_2 + c_3c_2 & b_3b_3 + c_3c_3 \end{pmatrix} \begin{bmatrix} V_1 \\ V_2 \\ V_3 \end{bmatrix} \tag{6.18}$$

TABLE 6.1
Electrostatics/Electrokinetics Duality

Electrostatics	Electrokinetics
$\oint_{\Sigma} \vec{D} \cdot d\vec{S} = Q_i$	$\oint_{\Sigma} \vec{J} \cdot d\vec{S} = 0$
$\vec{E} = -\nabla V$	$\vec{E} = -\nabla V$
$\vec{D} = \varepsilon \vec{E}$	$\vec{J} = \sigma \vec{E}$
$\vec{D} \; (\text{C/m}^2)$	$\vec{J} \; (\text{A/m}^2)$
$\varepsilon \;\; (\text{F/m})$	$\sigma \;\; (\text{S/m})$

As the second member of the continuity equation (4.21) is null, there is no vector corresponding to (6.15), so that Equation 6.16 is reduced to

$$\sum_{e=1}^{NE} E_i^e = 0 \quad i = 1, 2, \ldots, NN \tag{6.19}$$

This can be expressed in matrix form as follows:

$$[G][V] = [0] \tag{6.20}$$

As discussed in Chapter 4, both systems (6.17) and (6.20) are not only dual but also singular. This singularity is eliminated with the introduction of the boundary conditions of the problem, as already discussed.

6.4 MAGNETOSTATICS IN AXISYMMETRIC GEOMETRY

Figure 6.4 shows a typical problem of magnetostatics in axisymmetric geometry. The conductors represented in Figure 6.4 are the cross sections of coils (or massive conductors), the center of which is the z-axis from the cylindrical coordinates system.

In plane magnetostatics, this phenomenon is governed by Maxwell's second equation, given by

$$\oint_C \vec{H} \cdot d\vec{l} = \int_S \vec{J} \cdot d\vec{S} \tag{6.21}$$

On axisymmetric geometry, the current density vector and the magnetic potential vector are such that

$$\vec{J} = J(r,z)\vec{u}_\varphi \quad \text{and} \quad \vec{A} = A(r,z)\vec{u}_\varphi$$

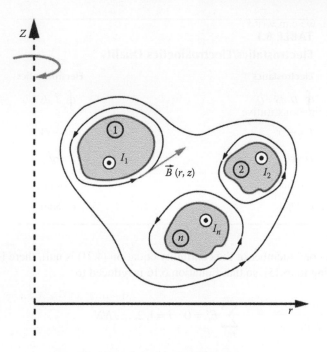

FIGURE 6.4 Axisymmetric magnetostatics typical problem.

As $\vec{B} = \nabla \times \vec{A}$, it can be written in cylindrical coordinates as follows:

$$\vec{B} = -\frac{\partial A}{\partial z}\vec{u}_r + \frac{1}{r}\frac{\partial(rA)}{\partial r}\vec{u}_z \tag{6.22}$$

It is observed that in problems with cylindrical symmetry (or axisymmetric), the current density vector and the magnetic potential vector are perpendicular to the plane (r, z) and the magnetic field is in the plane (r, z).

The method to solve these problems consists in evaluating the circulation of $\vec{H}$ on closed paths with polygonal geometry, involving each node of the finite element mesh. These polygonal closed paths—like in the plane symmetry—are composed of straight segments that link the centroid to the middle of the edges of the elements.

Thus, the line integral

$$E_1^e = \int\limits_{POS} \vec{H} \cdot \vec{dl} \tag{6.23}$$

is a section of the $\vec{H}$ circulation on the polygonal line involving node (1) that belongs to element (e).

Therefore, we can write

$$\sum_{e=1}^{NE} E_i^e = \oint_{C_i} \vec{H} \cdot \vec{dl} \quad i = 1, 2, ..., NN \tag{6.24}$$

Applying the constitutive relation $\vec{H} = v\vec{B}$, we get

$$\vec{H} = -v\frac{\partial A}{\partial z}\vec{u}_r + \frac{v}{r}\frac{\partial(rA)}{\partial r}\vec{u}_z \tag{6.25}$$

Defining the modified magnetic potential as $\varphi = rA$, the expression (6.25) can be rewritten as follows:

$$\vec{H} = -\frac{v}{r}\frac{\partial\varphi}{\partial z}\vec{u}_r + \frac{v}{r}\frac{\partial\varphi}{\partial r}\vec{u}_z \tag{6.26}$$

At this point, it is important to observe that on the finite element's domain we can consider $r \approx cte$, because its dimensions are smaller than those of the entire domain under study. With that, we can make the following approximation:

$$\vec{H} = -\frac{v}{r_b}\frac{\partial\varphi}{\partial z}\vec{u}_r + \frac{v}{r_b}\frac{\partial\varphi}{\partial r}\vec{u}_z \tag{6.27}$$

where

$$r_b = (r_1 + r_2 + r_3)/3$$

is the radius of finite element's centroid.

As φ presents the same continuity properties as the magnetic potential vector, we can estimate its value at any point inside the element through a linear interpolation of its values on the nodes, that is,

$$\varphi(r, z) = N_1\varphi_1 + N_2\varphi_2 + N_3\varphi_3 \tag{6.28}$$

where

$$N_i = \frac{1}{2\Delta}(a_i + b_i r + c_i z) \quad i = 1, 2, 3 \tag{6.29}$$

Applying this interpolation in Equation 6.27, we get

$$H_r = -\frac{v}{r_b}\frac{\partial\varphi}{\partial z} = -\frac{v}{2r_b\Delta}(c_1V_1 + c_2V_2 + c_3V_3)$$

$$H_z = \frac{v}{r_b}\frac{\partial\varphi}{\partial r} = \frac{v}{2r_b\Delta}(b_1V_1 + b_2V_2 + b_3V_3) \tag{6.30}$$

As $\vec{H}$ remains constant inside the element, it is possible to evaluate the line integral on both PO and OS directly by integration over the OS segment. Therefore, we can write

$$E_1^e = \int_{\overline{POS}} \vec{H} \cdot d\vec{l} = \vec{H} \cdot (S - P) \tag{6.31}$$

where

$$(S - P) = (r_s - r_p)\vec{u}_r + (z_s - z_p)(\vec{u}_z)$$

or

$$(S - P) = \frac{c_1}{2}\vec{u}_r - \frac{b_1}{2}\vec{u}_z$$

so that

$$E_1^e = \int_{\overline{POS}} \vec{H} \cdot d\vec{l} = -\frac{\nu}{4r_b\Delta}[(b_1b_1 + c_1c_1)\varphi_1 + (b_1b_2 + c_1c_2)\varphi_2 + (b_1b_3 + c_1c_3)\varphi_3] \tag{6.32}$$

Following a similar procedure for GOP and SOG sections, we obtain the line integrations sections that involve nodes (2) and (3), respectively. After some mathematical manipulation, we have

$$E_2^e = \int_{\overline{GOP}} \vec{H} \cdot d\vec{l} = -\frac{\nu}{4r_b\Delta}[(b_2b_1 + c_2c_1)\varphi_1 + (b_2b_2 + c_2c_2)\varphi_2 + (b_2b_3 + c_2c_3)\varphi_3] \tag{6.33}$$

$$E_3^e = \int_{\overline{SOG}} \vec{H} \cdot d\vec{l} = -\frac{\nu}{4r_b\Delta}[(b_3b_1 + c_3c_1)\varphi_1 + (b_3b_2 + c_3c_2)\varphi_2 + (b_3b_3 + c_3c_3)\varphi_3] \tag{6.34}$$

The results obtained in Equations 6.32 through 6.34 can be expressed in matrix form as follows:

$$\begin{bmatrix} E_1^e \\ E_2^e \\ E_3^e \end{bmatrix} = \frac{\nu}{4r_b\Delta} \begin{pmatrix} b_1b_1 + c_1c_1 & b_1b_2 + c_1c_2 & b_1b_3 + c_1c_3 \\ b_2b_1 + c_2c_1 & b_2b_2 + c_2c_2 & b_2b_3 + c_2c_3 \\ b_3b_1 + c_3c_1 & b_3b_2 + c_3c_2 & b_3b_3 + c_3c_3 \end{pmatrix} \begin{bmatrix} \varphi_1 \\ \varphi_2 \\ \varphi_3 \end{bmatrix} \tag{6.35}$$

The negative sign of the expression (6.35) was suppressed for simplicity, but that does not affect the final result. The vector from the second member of Maxwell's second equation will also have its sign changed, as we will see next.

6.4.1 CONCATENATED CURRENT WITH THE POLYGONAL LINE

The linkage current with the polygonal line involving each node of the finite element's mesh is evaluated by portions. Thus, for the portion of the polygon involving node (1), the total linkage current is that crossing the surface delimited by the polygon "1POS1" from the generic finite element, as shown in Figure 6.5.

This linkage current is calculated from the integral

$$I_1^e = \int_{1POS1} \vec{J} \cdot d\vec{S}$$ (6.36)

where

$$\vec{J} = J\vec{u}_\varphi$$

can be considered constant inside the element. For this reason, the expression (6.36) is written as follows:

$$I_1^e = \int_{1POS1} \vec{J} \cdot d\vec{S} = \vec{J} \cdot \Delta\vec{S}$$ (6.37)

By the adopted clockwise orientation on the circulation, the $\Delta\vec{S}$ vector is oriented on the opposite sense from the axis φ (note that the orientation sequence of the coordinated axis in cylindrical coordinates is r, φ, z). Therefore, this can be expressed in the following form:

$$\Delta\vec{S} = \Delta S(-\vec{u}_\varphi)$$

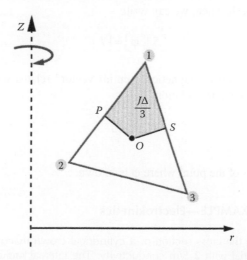

FIGURE 6.5 Portion of the concatenated current with the polygonal line involving node (1).

where ΔS is the hatched area on the polygon "1POS1," which is equivalent to one-third of the total area of the element, results

$$I_1^e = \int\limits_{1POS1} \vec{J} \cdot d\vec{S} = -J \frac{\Delta}{3}$$

As the other areas related to nodes (2) and (3) are identical to area "1POS1," we get

$$I_1^e = I_2^e = I_3^e = -J \frac{\Delta}{3}$$

As Equation (6.38) shows, these results can be represented as a matrix column in which negative sign is suppressed, matching with the signs exchange performed in Equation 6.35:

$$\begin{bmatrix} I_1^e \\ I_2^e \\ I_3^e \end{bmatrix} = \begin{bmatrix} J\Delta/3 \\ J\Delta/3 \\ J\Delta/3 \end{bmatrix} \tag{6.38}$$

Finally, the application of Maxwell's second equation on a polygonal line involving the generic node (i) can be written as follows:

$$\sum_{e=1}^{NE} E_i^e = \sum_{e=1}^{NE} I_i^e \quad i = 1, 2, \ldots, NN \tag{6.39}$$

This expression also generates a system of NN equations to NN unknowns, where the unknowns are the component $\vec{u}_\phi$ of the "modified magnetic potentials vector" of the finite element mesh. Then, we can write

$$[S][\varphi] = [I] \tag{6.40}$$

Note that the "modified magnetic potential vector" relates with the "magnetic potential vector" through the relation

$$A = \frac{\varphi}{r} \tag{6.41}$$

where r is the radius of the point where φ is known.

NUMERICAL EXAMPLE—Electrokinetics

Figure 6.6 shows the cross section of a cylindrical crown manufactured with a conducting material with 2 S/m conductivity. The internal radius of the crown has 1 m and the external radius 3 m. A voltage of 100 V is applied between the

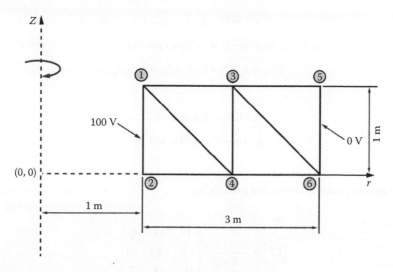

FIGURE 6.6   Cross section of a conductor cylindrical crown.

internal and the external face of the conductor, as shown in the figure. Let us calculate the potential of the nodes from the finite elements' mesh, the potentials of which are not known, and the electric field on each one of the elements.

Solution

The first step consists in identifying the type of problem. As it is about evaluating the potentials and the electric field in a conducting medium, we are dealing with an electrokinetics problem.

The following step is the assembly of the equations system that start with the matrixes associated to each mesh's element evaluated previously.

Thus, for each finite element, we will formulate a table with the following information:

Information for element (1)

Local numbering	1	2	3
Global numbering	1	2	4
Coordinate r (m)	1	1	2
Coordinate z (m)	1	0	0

The coefficients for these elements will be given by

$$b_1 = z_2 - z_3 = 0 - 0 \rightarrow b_1 = 0$$

$$b_2 = z_3 - z_1 = 0 - 1 \rightarrow b_2 = -1$$

$$b_3 = z_1 - z_2 = 1 - 0 \rightarrow b_3 = 1$$

$$c_1 = r_3 - r_2 = 2 - 1 \rightarrow c_1 = 1$$

$$c_2 = r_1 - r_3 = 1 - 2 \rightarrow c_2 = -1$$

$$c_3 = r_2 - r_1 = 1 - 1 \rightarrow c_3 = 0$$

$$\Delta = (b_1 c_2 - b_2 c_1)/2 = 1/2$$

$$r_b = (r_1 + r_2 + r_3)/3 = 4/3$$

Element's matrix—extracted from (6.18)

$$\left[G^1\right] = \frac{2\pi(4/3).2}{4(1/2)} \begin{pmatrix} 1 & -1 & 0 \\ -1 & 2 & -1 \\ 0 & -1 & 1 \end{pmatrix}$$

or

$$\left[G^1\right] = \frac{2\pi}{3} \begin{pmatrix} 1 & -1 & 0 \\ -1 & 2 & -1 \\ 0 & -1 & 1 \end{pmatrix}$$

The actions vector in the electrokinetics case is null and for this reason it is not calculated.

Following the same procedure for the other elements, we obtain the following matrixes:

$$\left[G^2\right] = \frac{2\pi}{3} \begin{pmatrix} 5 & 0 & -5 \\ 0 & 5 & -5 \\ -5 & -5 & 10 \end{pmatrix}$$

$$\left[G^3\right] = \frac{2\pi}{3} \begin{pmatrix} 7 & -7 & 0 \\ -7 & 14 & -7 \\ 0 & -7 & 7 \end{pmatrix}$$

$$\left[G^4\right] = \frac{2\pi}{3} \begin{pmatrix} 8 & 0 & -8 \\ 0 & 8 & -8 \\ -8 & -8 & 16 \end{pmatrix}$$

Applying the assembly technique from the global matrix and introducing the boundary conditions, we obtain the following system of equations:

$$
\begin{bmatrix}
1 & 0 & 0 & 0 & 0 & 0 \\
0 & 1 & 0 & 0 & 0 & 0 \\
0 & 0 & 25 & -12 & 0 & 0 \\
0 & 0 & -12 & 23 & 0 & 0 \\
0 & 0 & 0 & 0 & 1 & 0 \\
0 & 0 & 0 & 0 & 0 & 1
\end{bmatrix}
\cdot
\begin{bmatrix}
V_1 \\
V_2 \\
V_3 \\
V_4 \\
V_5 \\
V_6
\end{bmatrix}
=
\begin{bmatrix}
100 \\
100 \\
500 \\
400 \\
0 \\
0
\end{bmatrix}
$$

the solution of which supplies the required values of the potentials:

$$
V_1 = V_2 = 100(V); \quad V_3 = 37,819(V); \quad V_4 = 37,123(V); \quad V_5 = V_6 = 0(V)
$$

The evaluation of the electric field is made element by element using the following expressions:

$$
E_r = -\frac{\partial V}{\partial r} = -\frac{1}{2\Delta}(b_1 V_1 + b_2 V_2 + b_3 V_3)
$$

$$
E_z = -\frac{\partial V}{\partial z} = -\frac{1}{2\Delta}(c_1 V_1 + c_2 V_2 + c_3 V_3)
$$

For element (1)

$$
E_r = -\frac{1}{2(1/2)}[(0).(100) + (-1).(100) + (1).(37,123)] = 62,877 \text{ V/m}
$$

$$
E_z = -\frac{1}{2(1/2)}[(1).(100) + (-1).(100) + 0.(37,123)] = 0 \text{ V/m}
$$

and for the other elements as follows.
Element (2):

$$
E_r = 62,821 \text{ V/m}
$$

$$
E_z = -0.696 \text{ V/m}
$$

Element (3):

$$
E_r = 37,123 \text{ V/m}
$$

$$
E_z = -0.696 \text{ V/m}
$$

Element (4):

$$E_r = 37,819 \text{ V/m}$$

$$E_z = 0 \text{ V/m}$$

We can compare the results obtained by this methodology with the theoretical one. From electromagnetism, it is demonstrated, that the electric field in any point on the interior of the conductor is given by the expression

$$\vec{E}(r) = \frac{\Delta V}{r \ln \dfrac{b}{a}} \vec{u}_r$$

where
ΔV is the voltage between the internal and external faces from the cylindrical crown, which in this case is 100 (V).
a and b are the crown's internal and external radii

Applying this expression on the element's centroid (1), we get

$$\vec{E}(r) = \frac{100}{\dfrac{4}{3} \ln \dfrac{3}{1}} \vec{u}_r = 68,27 \vec{u}_r \ (\text{V/m})$$

This result diverges 9% of the calculated value, which is acceptable due to the poor-quality finite element's mesh.
The electric field on direction (z) is null, which justifies its reduced values obtained for elements (2) and (3).

6.5   MAGNETODYNAMICS IN AXISYMMETRIC GEOMETRY

The equation that governs the magnetodynamic phenomenon, as discussed in Chapter 5, is given by

$$\oint_C \vec{H} \cdot d\vec{l} = \int_S \vec{J} \cdot d\vec{S} - \int_S \sigma \frac{\partial \vec{A}}{\partial t} \cdot d\vec{S} \tag{6.42}$$

Evaluating each term of the domain integrals from the finite element and summing all closed path contributions involving each mesh's node, we get

$$\sum_{e=1}^{NE} E_i^e = \sum_{e=1}^{NE} I_i^e - \sum_{e=1}^{NE} \sigma F_i^e \quad i = 1, \ldots, NN$$

or

$$\sum_{e=1}^{NE} \sigma F_i^e + \sum_{e=1}^{NE} E_i^e = \sum_{e=1}^{NE} I_i^e \quad i = 1, \ldots, NN \tag{6.43}$$

From this matrix equation, we only have to determine how to assemble the first term because the others were vastly discussed in our discussion on axisymmetric magnetostatics.

Concerning the Equation 6.42, the first term of the right-hand side is the current impressed by the source and the second one is associated to the eddy currents due to the time-dependent magnetic field.

The integral from eddy current related to the modified magnetic potential vector φ is written as follows:

$$\int_S \sigma \frac{\partial A}{\partial t} \cdot dS = \int_S \frac{\sigma}{r} \frac{\partial \varphi}{\partial t} \cdot dS$$

which is evaluated on polygons "1POS1," 2GOP2," and "3SOG3," giving the following actions vector:

$$\begin{bmatrix} F_1^e \\ F_2^e \\ F_3^e \end{bmatrix} = \frac{\sigma \Delta}{12 r_b} \begin{pmatrix} 2 & 1 & 1 \\ 1 & 2 & 1 \\ 1 & 1 & 2 \end{pmatrix} \begin{bmatrix} \dot{\varphi}_1 \\ \dot{\varphi}_2 \\ \dot{\varphi}_3 \end{bmatrix} \tag{6.44}$$

Therefore, applying (6.41) to closed paths involving a generic node from the mesh gives

$$\sum_{e=1}^{NE} \sigma F_i^e + \sum_{e=1}^{NE} E_i^e = \sum_{e=1}^{NE} I_i^e \quad i = 1, \dots, NN \tag{6.45}$$

This can be globally represented in matrix form as follows:

$$\begin{bmatrix} C \end{bmatrix}\begin{bmatrix} \dot{\varphi} \end{bmatrix} + \begin{bmatrix} S \end{bmatrix}\begin{bmatrix} \varphi \end{bmatrix} = \begin{bmatrix} I \end{bmatrix} \tag{6.46}$$

which for each finite element is given by

$$C_{ij} = \begin{cases} \dfrac{\sigma \Delta}{6 r_b} & \text{for } i = j \\[3mm] \dfrac{\sigma \Delta}{12 r_b} & \text{for } i \neq j \end{cases} \tag{6.47}$$

$$S_{ij} = \frac{\nu}{4 r_b \Delta}(b_i b_j + c_i c_j) \tag{6.48}$$

$$I_j = J_f \frac{\Delta}{3} \tag{6.49}$$

The same assumptions made over the introduction of both the boundary and initial conditions related to the equations' system resolution (5.9) are also valid for the system (6.46).

In the case of a sinusoidal time-dependent magnetic field, the equations system is solved using a complex notation given by

$$\left[\hat{S}\right]\cdot\left[\hat{\phi}\right]=\left[\hat{I}\right] \tag{6.50}$$

where

$$\left[\hat{S}\right]=\left[S\right]+j\omega\left[C\right] \tag{6.51}$$

and $\left[\hat{\phi}\right]$ and $\left[\hat{I}\right]$ are column vectors where each term corresponds to the phasor (magnitude and phase) of the corresponding quantities.

7 Finite Element Method for High Frequency

7.1 INTRODUCTION

When fast, time-dependent electric and magnetic fields are present, their most important characteristic is propagation, which is typical of electromagnetic waves. As propagation in free space is three dimensional, its mathematical formulation is complex and beyond this book's scope. However, an interesting two-dimensional application is used in wave guide studies, because some peculiarities, discussed in this chapter, of wave propagation in a wave guide make possible its analysis through a two-dimensional formulation [48,66].

A wave guide is a device constituted by a piping structure, with metal walls, designed for guiding a wave through a short path in order to reproduce the signal sent at the other end with minimum losses.

Figure 7.1 shows a generic wave guide, the axial axis of which is coincident to the z-axis of the plane coordinates system.

Propagation in metal piping depends on certain conditions; for example, the most common occurs in coaxial cables, which, due to the current flowing in both the internal and external conductors, produce electromagnetic wave propagation on the dielectric without any component of the electric field in the propagation direction (*z direction*). This type of wave is called a transverse electromagnetic wave (TEM). Removing the internal conductor, we will fall over the classic standard of wave guide. As it is not possible to promote the electric current circulation on the internal conductor, as it does not exist, we have to establish a displacement current, that is, a high variation, time-dependent electric field with appropriate frequency and direction, because the guide dimensions impose constraints due to the boundary conditions, which shall be imposed on the guide walls. Several books on electromagnetism deal with this subject in detail and we will not discuss this here.

Electromagnetic wave propagation inside a wave guide can occur in two different ways. The first one is where there is a longitudinal electric field, that is, in the z direction, which is the propagation direction. In this case, it is not possible for a magnetic field to be lined with the electric field. This type of wave propagation is called transverse magnetic (TM) wave.

Another possibility is the inverse of the prior one, that is, the existence of a longitudinal magnetic field with a null longitudinal electric field. The propagation in this case is called transverse-electric (TE) wave. The existence of an eventual internal conductor, as in the case of the coaxial cable, besides the possibility of the existence of a TEM wave, this geometry also supports a TE or TM wave.

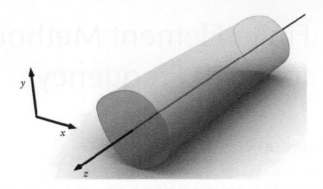

FIGURE 7.1   Wave guide.

Our objective is to present the mathematical methodology for obtaining the electric or magnetic field distribution on the cross section of the wave guide based on a 2D finite element method.

7.2   SPECIAL FEATURES OF MAXWELL'S EQUATIONS

When both the electric and magnetic fields are a sinusoidal time-dependent high frequency signals, the phenomenon is called time-harmonic state and propagation occurs. Supposing that the propagation medium is air, that is, the wave guide is hollow, both the first and second differential forms of Maxwell's equations can be written in complex notation as follows:

$$\nabla \times \hat{\vec{E}} = -j\omega\mu\hat{\vec{H}}$$

$$\nabla \times \hat{\vec{H}} = j\omega\varepsilon\hat{\vec{E}}$$

(7.1)

where $\hat{\vec{E}}$ and $\hat{\vec{H}}$ are, respectively, the complex electric and magnetic field vectors.

Due to this wave propagation characteristic of the wave guide, these fields are such that

$$\hat{\vec{E}} = \hat{\vec{E}}(x,y)\cdot e^{j(\omega t - \beta z)} = (\hat{E}_x \vec{u}_x + \hat{E}_y \vec{u}_y + \hat{E}_z \vec{u}_z)\cdot e^{j(\omega t - \beta z)}$$

$$\hat{\vec{H}} = \hat{\vec{H}}(x,y)\cdot e^{j(\omega t - \beta z)} = (\hat{H}_x \vec{u}_x + \hat{H}_y \vec{u}_y + \hat{H}_z \vec{u}_z)\cdot e^{j(\omega t - \beta z)}$$

(7.2)

Each complex vector component, $\hat{\vec{E}}$ and $\hat{\vec{H}}$, represents the magnitude and the phase of the $e^{j(\omega t - \beta z)}$ respective component, which depends exclusively on the coordinates x,y.

Note that the term characterizes that all the complex vector components are sine waves that propagate in the direction $z > 0$, which is the wave guide's axis, at a propagation speed given by $v = \omega/\beta$.

This results in the following equation:

$$\frac{\partial \hat{\vec{E}}}{\partial z} = -j\beta\hat{\vec{E}} \quad \text{and} \quad \frac{\partial \hat{\vec{H}}}{\partial z} = -j\beta\hat{\vec{H}} \tag{7.3}$$

Expanding the Equation 7.1 using these considerations gives

$$\nabla \times \hat{\vec{E}} = -j\omega\mu\hat{\vec{H}} \begin{cases} \dfrac{\partial \hat{E}_z}{\partial y} - \dfrac{\partial \hat{E}_y}{\partial z} = -j\omega\mu\hat{H}_x \\[2mm] \dfrac{\partial \hat{E}_x}{\partial z} - \dfrac{\partial \hat{E}_z}{\partial x} = -j\omega\mu\hat{H}_y \\[2mm] \dfrac{\partial \hat{E}_y}{\partial x} - \dfrac{\partial \hat{E}_x}{\partial y} = -j\omega\mu\hat{H}_z \end{cases} \tag{7.4}$$

$$\nabla \times \hat{\vec{H}} = j\omega\varepsilon\hat{\vec{E}} \begin{cases} \dfrac{\partial \hat{H}_z}{\partial y} - \dfrac{\partial \hat{H}_y}{\partial z} = j\omega\varepsilon\hat{E}_x \\[2mm] \dfrac{\partial \hat{H}_x}{\partial z} - \dfrac{\partial \hat{H}_z}{\partial x} = j\omega\varepsilon\hat{E}_y \\[2mm] \dfrac{\partial \hat{H}_y}{\partial x} - \dfrac{\partial \hat{H}_x}{\partial y} = j\omega\varepsilon\hat{E}_z \end{cases} \tag{7.5}$$

Applying (7.3) in Equations 7.4 and 7.5, the prior equations are rewritten as follows:

$$\nabla \times \hat{\vec{E}} = -j\omega\mu\hat{\vec{H}} \begin{cases} \dfrac{\partial \hat{E}_z}{\partial y} + j\beta\hat{E}_y = -j\omega\mu\hat{H}_x \\[2mm] -j\beta\hat{E}_x - \dfrac{\partial \hat{E}_z}{\partial x} = -j\omega\mu\hat{H}_y \\[2mm] \dfrac{\partial \hat{E}_y}{\partial x} - \dfrac{\partial \hat{E}_x}{\partial y} = -j\omega\mu\hat{H}_z \end{cases} \tag{7.6}$$

$$\nabla \times \hat{\vec{H}} = j\omega\varepsilon\hat{\vec{E}} \begin{cases} \dfrac{\partial \hat{H}_z}{\partial y} + j\beta\hat{H}_y = j\omega\varepsilon\hat{E}_x \\[2mm] -j\beta\hat{H}_x - \dfrac{\partial \hat{H}_z}{\partial x} = j\omega\varepsilon\hat{E}_y \\[2mm] \dfrac{\partial \hat{H}_y}{\partial x} - \dfrac{\partial \hat{H}_x}{\partial y} = j\omega\varepsilon\hat{E}_z \end{cases} \tag{7.7}$$

From (7.6) and (7.7), all components in directions (x) and (y) can be expressed based on the direction of components in the (z) direction. As an example, for obtaining the component (x) from the magnetic vector (H_x), we should isolate (E_y) from the second system's equation (7.7) and replace it in the first system's equation (7.6), and then isolate (H_x). Following a similar procedure for the others components results in the following:

$$\hat{H}_x = \frac{j}{k^2 - \beta^2}\left(\omega\varepsilon\frac{\partial\hat{E}_z}{\partial y} - \beta\frac{\partial\hat{H}_z}{\partial x}\right); \quad \hat{H}_y = \frac{-j}{k^2 - \beta^2}\left(\omega\varepsilon\frac{\partial\hat{E}_z}{\partial x} + \beta\frac{\partial\hat{H}_z}{\partial y}\right) \quad (7.8)$$

$$\hat{E}_x = \frac{-j}{k^2 - \beta^2}\left(\beta\frac{\partial\hat{E}_z}{\partial x} + \omega\mu\frac{\partial\hat{H}_z}{\partial y}\right); \quad \hat{E}_y = \frac{j}{k^2 - \beta^2}\left(-\beta\frac{\partial\hat{E}_z}{\partial y} + \omega\mu\frac{\partial\hat{H}_z}{\partial x}\right) \quad (7.9)$$

where

$$k^2 = \omega^2\mu\varepsilon$$

7.3 TM MODE; $H_z = 0$

In a TM propagation

$$\hat{E}_z \neq 0 \quad \text{with } \hat{H}_z = 0$$

Under these conditions, Equations 7.8 and 7.9 are rewritten as follows:

$$\hat{H}_x = \frac{j\omega\varepsilon}{k^2 - \beta^2}\frac{\partial\hat{E}_z}{\partial y}; \quad \hat{H}_y = \frac{-j\omega\varepsilon}{k^2 - \beta^2}\frac{\partial\hat{E}_z}{\partial x} \quad (7.10)$$

$$\hat{E}_x = \frac{-j\beta}{k^2 - \beta^2}\frac{\partial\hat{E}_z}{\partial x}; \quad \hat{E}_y = \frac{-j\beta}{k^2 - \beta^2}\frac{\partial\hat{E}_z}{\partial y} \quad (7.11)$$

It is observed that all the components from both the magnetic and electric fields are expressed based on the (z) component of the electric field. Thus, let us implement the methodology for establishing $\hat{E}_z$ through the finite element method. Once this component of the electric field is obtained, the wave guide can be totally characterized.

Figure 7.2 shows the cross section of a wave guide discretized with first-order triangular elements. The idea is always the same; let us apply Maxwell's second equation to the polygonal line constituted by straight segments that link the centroid to the middle of the sides of the element.

This procedure is similar to those used in previous studies. The difference now is that instead of calculating the (electric or magnetic) potential on each mesh's node, we calculate, directly, the (z) component of the electric field.

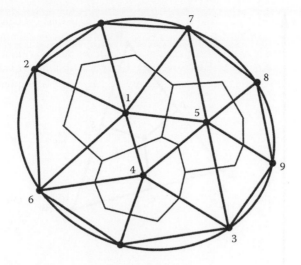

FIGURE 7.2 Wave guide—mesh of finite element.

The integral form of Maxwell's second equation written in the complex notation is given by

$$\oint_C \hat{\vec{H}} d\vec{S} = \int_S \left(\hat{\vec{J}} + j\omega \hat{\vec{D}} \right) \cdot d\vec{S} \tag{7.12}$$

The conduction current inside the guide does not exist due to the dielectric, and the propagation speed is not the speed of light. This is because the electromagnetic waves hitting the walls of the device produce shockwaves and cause delays. Applying the constitutive relation $D = \varepsilon E$, Equation 7.12 is reduced to

$$\oint_C \hat{\vec{H}} \cdot d\vec{l} = j\omega \int_S \hat{\vec{E}} \cdot d\vec{S} \tag{7.13}$$

The equations for the wave guide geometry are

$$\hat{\vec{H}} = \hat{H}_x \vec{u}_x + \hat{H}_y \vec{u}_y; \, \hat{\vec{E}} = \hat{E}_z \vec{u}_z$$

$$\vec{dl} = dx\vec{u}_x + dy\vec{u}_y; \, d\vec{S} = dS\vec{u}_z$$

so that

$$\oint_C \left(\hat{H}_x dx + \hat{H}_y dy \right) = j\omega\varepsilon \int_S \hat{E}_z \cdot d\vec{S} \tag{7.14}$$

with $\hat{H}_x$ and $\hat{H}_y$ given by Equation 7.10.

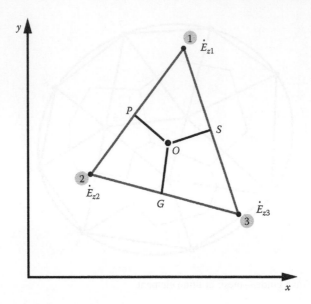

FIGURE 7.3   Generic finite element.

Equation 7.14 will be applied to polygonal sections with the same characteristics as those used in the other studies. Thus, inside each element, partial integrations of integrals, which compound both members of the Equation, can be calculated from the finite element's mesh.

First, let us establish the matrix expression, which is produced by the second member's expression. For this, let us consider the generic finite element shown in Figure 7.3.

Our objective is to evaluate the flux of $\hat{E}_z$ over the polygonal surface that denotes each node of the element.

Each finite element will contribute to a portion of this flux for three different surfaces involving nodes 1, 2, and 3. For the portion corresponding to the surface that involves node 1, it is necessary to evaluate the flux of $\hat{E}_z$ over the surface surrounded by the polygon "1POS1." As component z of the electric field vector can be considered linearly dependent inside the element, we can write

$$\hat{E}_z = N_1\hat{E}_{z1} + N_2\hat{E}_{z2} + N_3\hat{E}_{z3} \tag{7.15}$$

Let us estimate the average value of $\hat{E}_z$ over the surface "1POS1" from its values on four points of an irregular triangle:

1. Node 1: $\hat{E}_{z1}$

2. Node P: $E_P = \dfrac{1}{2}(\dot{E}_{z1} + \dot{E}_{z2})$

3. Node O: $\dot{E}_O = \frac{1}{3}(\dot{E}_{z1} + \dot{E}_{z2} + \dot{E}_{z3})$

4. Node S: $\dot{E}_S = \frac{1}{2}(\dot{E}_{z1} + \dot{E}_{z3})$

The average value of the electric field on the interior of this polygon to be considered is the value evaluated on the middle point of PS (point K) segment, the value of which is given by

$$\dot{E}_K = \frac{1}{2}(\dot{E}_P + \dot{E}_S) = \frac{1}{2}\left(\frac{\dot{E}_{z1} + \dot{E}_{z2}}{2} + \frac{\dot{E}_{z1} + \dot{E}_{z3}}{2}\right) = \frac{1}{2}\left(\frac{2\dot{E}_{z1} + \dot{E}_{z2} + \dot{E}_{z3}}{2}\right) \quad (7.16)$$

resulting in

$$F_1^e = \int_{1POS1} \hat{E}_z dS$$

which represents the flux of $\dot{E}_z$ over the surface delimited by polygon "1POS1" given by (Figure 7.4)

$$F_1^e = \frac{\Delta}{3}\hat{E}_{zk} = \frac{\Delta}{12}(2\hat{E}_{z1} + \hat{E}_{z2} + \hat{E}_{z3})$$

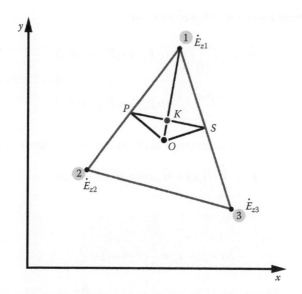

FIGURE 7.4 Average value of $\dot{E}_z$.

Following a similar procedure, for the other surfaces ("2GOP2" and "3SOG3")

$$F_2^e = \frac{\Delta}{12}(\hat{E}_{z1} + 2\hat{E}_{z2} + \hat{E}_{z3})$$

$$F_3^e = \frac{\Delta}{12}(\hat{E}_{z1} + \hat{E}_{z2} + 2\hat{E}_{z3})$$

These results can be represented in matrix form as follows:

$$\begin{bmatrix} F_1^e \\ F_2^e \\ F_3^e \end{bmatrix} = \frac{\Delta}{12}\begin{pmatrix} 2 & 1 & 1 \\ 1 & 2 & 1 \\ 1 & 1 & 2 \end{pmatrix}\begin{bmatrix} \hat{E}_{z1} \\ \hat{E}_{z2} \\ \hat{E}_{z3} \end{bmatrix} \tag{7.17}$$

The right-hand side of the circulation expressed in Equation 7.14 is applied to the polygon involving each mesh's node and, as in the previous example, it is also calculated by sections. Thus, on the part of the polygonal line which involves the node (1), we can write

$$E_1^e = \int_{POS} (\hat{H}_x dx + \hat{H}_y dy) \tag{7.18}$$

From Equation 7.10, we obtain

$$\hat{H}_x = \frac{j\omega\varepsilon}{k^2 - \beta^2}\frac{\partial \hat{E}_z}{\partial y}; \quad \hat{H}_y = \frac{-j\omega\varepsilon}{k^2 - \beta^2}\frac{\partial \hat{E}_z}{\partial x}$$

Therefore, from Equation 7.15, we can write

$$\hat{H}_x = \frac{j\omega\varepsilon}{k^2 - \beta^2}\frac{1}{2\Delta}(c_1\hat{E}_{z1} + c_2\hat{E}_{z2} + c_3\hat{E}_{z3})$$

$$\hat{H}_y = \frac{-j\omega\varepsilon}{k^2 - \beta^2}\frac{1}{2\Delta}(b_1\hat{E}_{z1} + b_2\hat{E}_{z2} + b_3\hat{E}_{z3})$$

Replacing these expressions obtained in (7.18), the section circulation results in

$$E_1^e = \int_{POS} (\hat{H}_x dx + \hat{H}_y dy) = \hat{H}_x \Delta x - \hat{H}_y \Delta y \tag{7.19}$$

where

$$\Delta x = x_s - x_p = \frac{1}{2}(x_3 - x_2) = \frac{c_1}{2}$$

$$\Delta y = y_p - y_s = \frac{1}{2}(y_2 - y_3) = \frac{b_1}{2}$$

Replacing $\hat{H}_x$ and $\hat{H}_y$ by their values in Equation 7.19, we get

$$E_1^e = \frac{j\omega\varepsilon}{k^2-\beta^2}\frac{1}{4\Delta}[(b_1b_1+c_1c_1)\hat{E}_{z1}+(b_1b_2+c_1c_2)\hat{E}_{z2}+(b_1b_3+c_1c_3)\hat{E}_{z3}]$$

Applying the same procedure for section circulations involving nodes (2) and (3), we get

$$E_2^e = \frac{j\omega\varepsilon}{k^2-\beta^2}\frac{1}{4\Delta}[(b_2b_1+c_2c_1)\hat{E}_{z1}+(b_2b_2+c_2c_2)\hat{E}_{z2}+(b_2b_3+c_2c_3)\hat{E}_{z3}]$$

$$E_3^e = \frac{j\omega\varepsilon}{k^2-\beta^2}\frac{1}{4\Delta}[(b_3b_1+c_3c_1)\hat{E}_{z1}+(b_3b_2+c_3c_2)\hat{E}_{z2}+(b_3b_3+c_3c_3)\hat{E}_{z3}]$$

The section circulations can be written in matrix from as follows:

$$\begin{bmatrix} E_1^e \\ E_2^e \\ E_3^e \end{bmatrix} = \frac{j\omega\varepsilon}{k^2-\beta^2}\frac{1}{4\Delta}\begin{pmatrix} b_1b_1+c_1c_1 & b_1b_2+c_1c_2 & b_1b_3+c_1c_3 \\ b_2b_1+c_2c_1 & b_2b_2+c_2c_2 & b_2b_3+c_2c_3 \\ b_3b_1+c_3c_1 & b_3b_2+c_3c_2 & b_3b_3+c_3c_3 \end{pmatrix}\begin{bmatrix} \hat{E}_{z1} \\ \hat{E}_{z2} \\ \hat{E}_{z3} \end{bmatrix} \qquad (7.20)$$

Following this, Maxwell's second equation (7.14) can be applied to the polygonal line involving the node (i) as follows:

$$\sum_{e=1}^{NE} E_i^e = \sum_{e=1}^{NE} j\omega\varepsilon F_i^e \quad i=1,...,NN \qquad (7.21)$$

This can be globally represented in matrix form as follows:

$$\begin{bmatrix} S_{TM} \end{bmatrix}\begin{bmatrix} \hat{E}_z \end{bmatrix} = \begin{bmatrix} T_{TM} \end{bmatrix}\begin{bmatrix} \hat{E}_z \end{bmatrix} \qquad (7.22)$$

The elements of which are given by

$$T_{TMij} = \begin{cases} \dfrac{j\omega\varepsilon\Delta}{6} & \text{for } i=j \\[3mm] \dfrac{j\omega\varepsilon\Delta}{12} & \text{for } i\neq j \end{cases} \qquad (7.23)$$

$$S_{TMij} = \frac{j\omega\varepsilon}{k^2-\beta^2}\frac{1}{4\Delta}(b_ib_j+c_ic_j) \qquad (7.24)$$

7.4   TE MODE; $E_z = 0$

In a TE propagation

$$\hat{E}_z = 0 \quad \text{with } \hat{H}_z \neq 0$$

Under these conditions, Equations 7.8 and 7.9 are rewritten as follows:

$$\hat{H}_x = -\frac{j}{k^2 - \beta^2}\beta\frac{\partial \hat{H}_z}{\partial x}; \quad \hat{H}_y = \frac{-j}{k^2 - \beta^2}\beta\frac{\partial \hat{H}_z}{\partial y} \tag{7.25}$$

$$\hat{E}_x = \frac{-j}{k^2 - \beta^2}\omega\mu\frac{\partial \hat{H}_z}{\partial y}; \quad \hat{E}_y = \frac{j}{k^2 - \beta^2}\omega\mu\frac{\partial \hat{H}_z}{\partial x} \tag{7.26}$$

It can be observed that all components of both the magnetic and electric fields are expressed based on the (z) component of the magnetic field. Thus, as done previously, let us implement the methodology for establishing $\hat{H}_z$ through the finite element method, and once this component of the magnetic intensity vector is obtained, the wave guide can be totally characterized.

The Maxwell's equation to be used in this case is the first equation in its integral form, which, using complex notation and the constitutive relation $B = \mu H$, can be written as follows:

$$\oint_C \hat{\vec{E}} \cdot d\vec{l} = -j\omega\mu \int_S \hat{\vec{H}} \cdot d\vec{S} \tag{7.27}$$

and

$$\hat{\vec{E}} = \hat{E}_x\vec{u}_x + \hat{E}_y\vec{u}_y; \; \hat{\vec{H}} = \hat{H}_z\vec{u}_z$$

$$d\vec{l} = dx\vec{u}_x + dy\vec{u}_y; \; d\vec{S} = dS\vec{u}_z$$

which results in

$$\oint_C \left(\hat{E}_x dx + \hat{E}_y dy\right) = -j\omega\varepsilon \int_S \hat{H}_z \cdot d\vec{S} \tag{7.28}$$

with $\hat{E}_x$ and $\hat{E}_y$ taken from Equation 7.26.

Equation 7.28 is now applied to the polygonal line with the same characteristics used in the other studies:

$$\hat{H}_z = N_1\hat{H}_{z1} + N_2\hat{H}_{z2} + N_3\hat{H}_{z3} \tag{7.29}$$

The average value of $\hat{H}_z$ over the surface "1POS1" from its values on four points of an irregular triangle is, as in the previous case, given by

$$\hat{H}_{zk} = \frac{1}{4}(2\hat{H}_{z1} + \hat{H}_{z2} + \hat{H}_{z3}) \tag{7.30}$$

so that the second member's integral (7.28) becomes

$$F_1^e = \int_{1POS1} \hat{H}_z dS$$

which represents the flux of $\hat{H}_z$ over the surface delimited by polygon "1POS1" given by

$$F_1^e = \frac{\Delta}{3}\hat{H}_{zk} = \frac{\Delta}{12}(2\hat{H}_{z1} + \hat{H}_{z2} + \hat{H}_{z3})$$

Following the same procedure for the other surfaces ("2GOP2" and "3SOG3"), we get

$$F_2^e = \frac{\Delta}{12}(\hat{H}_{z1} + 2\hat{H}_{z2} + \hat{H}_{z3})$$

$$F_3^e = \frac{\Delta}{12}(\hat{H}_{z1} + \hat{H}_{z2} + 2\hat{H}_{z3})$$

This result can be written in matrix form as follows:

$$\begin{bmatrix} F_1^e \\ F_2^e \\ F_3^e \end{bmatrix} = \frac{\Delta}{12}\begin{pmatrix} 2 & 1 & 1 \\ 1 & 2 & 1 \\ 1 & 1 & 2 \end{pmatrix}\begin{bmatrix} \hat{H}_{z1} \\ \hat{H}_{z2} \\ \hat{H}_{z3} \end{bmatrix} \tag{7.31}$$

The first term's circulation is applied to the polygonal line involving each mesh's node; it is also calculated by sections. Thus, for the section of the polygonal line that involves node (1), we can write

$$E_1^e = \int_{POS} (\hat{E}_x dx + \hat{E}_x dy) \tag{7.32}$$

From Equation 7.26, we obtain

$$\hat{E}_x = \frac{-j}{k^2 - \beta^2}\omega\mu\frac{\partial\hat{H}_z}{\partial y}; \quad \hat{E}_y = \frac{j}{k^2 - \beta^2}\omega\mu\frac{\partial\hat{H}_z}{\partial x}$$

Hence, from Equation 7.29, we can write

$$\hat{E}_x = \frac{-j}{k^2 - \beta^2} \omega\mu \frac{1}{2\Delta}(c_1\hat{H}_{z1} + c_2\hat{H}_{z2} + c_3\hat{H}_{z3})$$

$$\hat{E}_y = \frac{j}{k^2 - \beta^2} \omega\mu \frac{1}{2\Delta}(b_1\hat{H}_{z1} + b_2\hat{H}_{z2} + b_3\hat{H}_{z3})$$

Replacing these results in Equation 7.32, the section circulation is calculated as follows:

$$E_1^e = \int_{POS} (\hat{E}_x dx + \hat{E}_y dy) = \hat{E}_x \Delta x - \hat{E}_y \Delta y \qquad (7.33)$$

where

$$\Delta x = x_s - x_p = \frac{1}{2}(x_3 - x_2) = \frac{c_1}{2}$$

$$\Delta y = y_p - y_s = \frac{1}{2}(y_2 - y_3) = \frac{b_1}{2}$$

Replacing $\hat{E}_x$ and $\hat{E}_y$ in Equation 7.33 results in

$$E_1^e = \frac{-j\omega\mu}{k^2 - \beta^2} \frac{1}{4\Delta}[(b_1b_1 + c_1c_1)\hat{H}_{z1} + (b_1b_2 + c_1c_2)\hat{H}_{z2} + (b_1b_3 + c_1c_3)\hat{H}_{z3}]$$

Applying the same procedure for section circulations involving nodes (2) and (3), we obtain

$$E_2^e = \frac{-j\omega\mu}{k^2 - \beta^2} \frac{1}{4\Delta}[(b_2b_1 + c_2c_1)\hat{H}_{z1} + (b_2b_2 + c_2c_2)\hat{H}_{z2} + (b_2b_3 + c_2c_3)\hat{H}_{z3}]$$

$$E_3^e = \frac{-j\omega\mu}{k^2 - \beta^2} \frac{1}{4\Delta}[(b_3b_1 + c_3c_1)\hat{H}_{z1} + (b_3b_2 + c_3c_2)\hat{H}_{z2} + (b_3b_3 + c_3c_3)\hat{H}_{z3}]$$

These section circulations can be written in matrix form as follows:

$$\begin{bmatrix} E_1^e \\ E_2^e \\ E_3^e \end{bmatrix} = \frac{-j\omega\mu}{k^2 - \beta^2} \frac{1}{4\Delta} \begin{pmatrix} b_1b_1 + c_1c_1 & b_1b_2 + c_1c_2 & b_1b_3 + c_1c_3 \\ b_2b_1 + c_2c_1 & b_2b_2 + c_2c_2 & b_2b_3 + c_2c_3 \\ b_3b_1 + c_3c_1 & b_3b_2 + c_3c_2 & b_3b_3 + c_3c_3 \end{pmatrix} \begin{bmatrix} \hat{H}_{z1} \\ \hat{H}_{z2} \\ \hat{H}_{z3} \end{bmatrix} \qquad (7.34)$$

Following this, the application of Maxwell's first equation (7.32) to the polygonal line involving node (i) can be written as follows:

$$\sum_{e=1}^{NE} E_i^e = -\sum_{e=1}^{NE} j\omega\mu F_i^e \quad i=1,\ldots,NN$$

or

$$-\sum_{e=1}^{NE} E_i^e = \sum_{e=1}^{NE} j\omega\mu F_i^e \quad i=1,\ldots,NN \tag{7.35}$$

This can be globally written in matrix form as follows:

$$\left[S_{TE}\right]\left[\hat{H}_z\right] = \left[T_{TE}\right]\left[\hat{H}_z\right] \tag{7.36}$$

the elements of which are given by

$$T_{TEij} = \begin{cases} \dfrac{j\omega\mu\Delta}{6} & \text{for } i=j \\[2mm] \dfrac{j\omega\mu\Delta}{12} & \text{for } i \neq j \end{cases} \tag{7.37}$$

$$S_{TEij} = \frac{j\omega\mu}{k^2-\beta^2}\frac{1}{4\Delta}(b_i b_j + c_i c_j) \tag{7.38}$$

Supposing the dielectric of the wave guide is both linear and isotropic, the matrix Equations 7.24 and 7.36 can be simplified to the general form

$$\left[S\right]\left[\hat{\Phi}\right] = \bar{k}^2\left[T\right]\left[\hat{\Phi}\right] \tag{7.39}$$

where

$$T_{ij} = \begin{cases} \dfrac{\Delta}{6} & \text{for } i=j \\[2mm] \dfrac{\Delta}{12} & \text{for } i \neq j \end{cases}$$

$$S_{ij} = \frac{1}{4\Delta}(b_i b_j + c_i c_j)$$

$$\bar{k}^2 = k^2 - \beta^2$$

$$\hat{\Phi} = \begin{cases} \hat{E}_z \text{ for TM wave} \\ \hat{H}_z \text{ for TE wave} \end{cases}$$

Note that if $\bar{k} = 0$, there is no propagation, and a frequency called "cutting frequency" occurs such that

$$\bar{k}^2 = k^2 - \beta^2 = 0$$

or

$$k = \beta$$

Multiplying Equation 7.39 by $[T]^{-1}$ after some mathematical manipulation, we get

$$\left[[T]^{-1}[S] - \bar{k}^2 [I] \right] \cdot \left[\hat{\Phi} \right] = [0] \tag{7.40}$$

which results in

$$A = [T]^{-1}[S]; \quad \bar{k}^2 = \lambda \quad e \quad X = \left[\hat{\Phi} \right]$$

The standard form of the autovalues and autovectors problems is obtained as follows:

$$(A - \lambda I)X = 0 \tag{7.41}$$

where I is the identity matrix.

Several procedures allow us to evaluate the values of λ (called eigenvalues) and X (called eigenvectors) that satisfy Equation 7.41.

This methodology makes possible to evaluate the cutting frequencies, extracted from eigenvalues and the magnitudes of fields for each eigenvalue from the obtainment of its eigenvectors.

8 Three-Dimensional Finite Element Method

8.1 INTRODUCTION

Although 2D formalism of the finite element method (FEM) for electromagnetics is a powerful tool for analyzing several kinds of devices, we need to understand that 2D representation is not enough to characterize more complex devices. In that case, 3D modeling is preferred. We face such a situation when the devices are too small, such as nanodevices, or ones whose geometry is highly complex [60].

The difficulty in 3D FEM is in the modeler, because creating a 3D model of an object is not simple. Nowadays, several good modelers with enough features that satisfy any requirement specified for an FEM analysis are available. Therefore, the automatic generation of a good 3D discretization is still an important subject of research.

Similar to our discussions on 2D formulation, we will constrain our study here to a simple 3D element, such as a tetrahedron. The tetrahedron is the only element to which we can apply first-order approximation that enables us to introduce the FEM for electromagnetics directly from Maxwell's equations without using complex formulations.

8.2 SHAPE FUNCTIONS FOR A TETRAHEDRON ELEMENT

The discretization of the domain in a tetrahedron element should obey the same criteria used in 2D discretization, which are the following:

1. There should be only one material inside the element.
2. The density of elements should be bigger in the regions we expect a strong variation in the field.
3. The intersection of two tetrahedron elements should be a face, an edge, or a vertex.

Figure 8.1 shows a 3D domain subdivided into tetrahedron elements.

Figure 8.2 highlights a generic finite element extracted from the domain using two different ways of numbering:

1. *Local numbering*: The vertexes are numbered from 1 to 4 applying the right-hand rule in the anticlockwise sense arbitrary choosing the vertex 1.
2. *Global numbering*: This is the number given to the vertex during random numbering from 1 to NN. In the figure, the global numbering of the vertexes are (p), (q), (r), and (t).

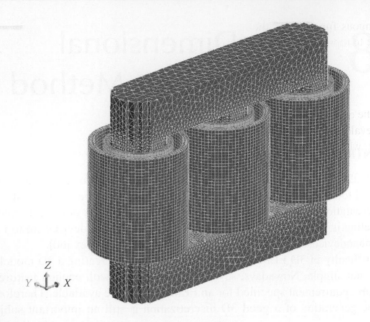

FIGURE 8.1   3D domain discretized in finite elements.

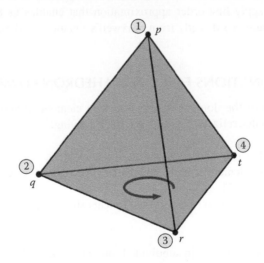

FIGURE 8.2   First-order tetrahedron element.

The FEM analysis aims to evaluate the nodal variable that can be a scalar function as the electric potential or a vector field one as the magnetic potential vector or both as we saw when we studied time-dependent electromagnetic field.

With that in mind, let us consider that the scalar function $f(x, y, z)$ represents the space variation of a physical quantity inside the element. Supposing that $f(x, y, z)$ is a continuous function, as the case in both electrical and magnetic potentials, it is possible to represent it through a first-order function as follows:

$$f(x,y,z) = \alpha_1 + \alpha_2 x + \alpha_3 y + \alpha_4 z \tag{8.1}$$

where the coefficients α_1, α_2, α_3, and α_4 should be determined.

For evaluating these, we need only write the function $f(x, y, z)$ at the nodes of the element, which gives

$$
\begin{aligned}
f_1 &= f(x_1, y_1, z_1) = \alpha_1 + \alpha_2 x_1 + \alpha_3 y_1 + \alpha_4 z_1 \\
f_2 &= f(x_2, y_2, z_2) = \alpha_1 + \alpha_2 x_2 + \alpha_3 y_2 + \alpha_4 z_2 \\
f_3 &= f(x_3, y_3, z_3) = \alpha_1 + \alpha_2 x_3 + \alpha_3 y_3 + \alpha_3 z_3 \\
f_4 &= f(x_4, y_4, z_4) = \alpha_1 + \alpha_2 x_4 + \alpha_3 y_4 + \alpha_4 z_4
\end{aligned}
\tag{8.2}
$$

Solving the 4×4-equation system in Equation 8.2, we obtain the coefficients we are looking for

$$\alpha_1 = \frac{1}{6V}(a_1 f_1 + a_2 f_2 + a_3 f_3 + a_4 f_4)$$

$$\alpha_2 = \frac{1}{6V}(b_1 f_1 + b_2 f_2 + b_3 f_3 + b_4 f_4)$$

$$\alpha_3 = \frac{1}{6V}(c_1 f_1 + c_2 f_2 + c_3 f_3 + c_4 f_4) \tag{8.3}$$

$$\alpha_4 = \frac{1}{6V}(d_1 f_1 + d_2 f_2 + d_3 f_3 + d_4 f_4)$$

where

$$
a_1 = \begin{vmatrix} x_2 & y_2 & z_2 \\ x_3 & y_3 & z_3 \\ x_4 & y_4 & z_4 \end{vmatrix}; \quad
b_1 = \begin{vmatrix} 1 & y_2 & z_2 \\ 1 & y_3 & z_3 \\ 1 & y_4 & z_4 \end{vmatrix}; \quad
c_1 = \begin{vmatrix} 1 & x_2 & z_2 \\ 1 & x_3 & z_3 \\ 1 & x_4 & z_4 \end{vmatrix} \quad \text{and} \quad
d_1 = \begin{vmatrix} 1 & x_2 & y_2 \\ 1 & x_3 & y_3 \\ 1 & x_4 & y_4 \end{vmatrix} (*)
$$

$$
6V = \det \begin{vmatrix} 1 & x_1 & y_1 & z_1 \\ 1 & x_2 & y_2 & z_2 \\ 1 & x_3 & y_3 & z_3 \\ 1 & x_4 & y_4 & z_4 \end{vmatrix}
$$

The other coefficients are obtained by cyclic rotation of the indexes, where V is the finite element's volume.

Replacing these coefficients in Equation 8.1, we obtain, after some mathematical manipulation, the following expression:

$$f(x,y,z) = N_1(x,y,z)f_1 + N_2(x,y,z)f_2 + N_3(x,y,z)f_3 + N_4(x,y,z)f_4 \qquad (8.4)$$

where

$$N_i(x,y,z) = \frac{1}{6V}(a_i + b_i x + c_i y + d_i z) \quad i = 1, 2, 3 \qquad (8.5)$$

As in the 2D case, the functions $N_1(x, y, z)$, $N_2(x, y, z)$, $N_3(x, y, z)$, and $N_4(x, y, z)$ are called "the shape functions of the first-order tetrahedron finite element." They also satisfy the following relation:

$$N_i(x_j, y_j, z_j) = \begin{cases} 1 & \text{if } i = j \\ \\ 0 & \text{if } i \neq j \end{cases} \qquad \text{and} \qquad \sum_{i=1}^{3} N_i(x,y,z) = 1$$

As we can see, the value of the ith function is (1) at node (i) and null at the other three nodes.

A compact form for representing the *first*-order interpolation function is

$$f(x,y,z) = \sum_{i=1}^{3} N_i(x,y,z)f_i \qquad (8.6)$$

Figure 8.3 shows a geometric interpretation of the shape function, which is the quotient between the tetrahedron composed by the point P and the base and the total volume of the tetrahedron.

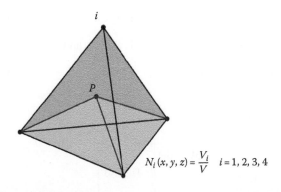

$$N_i(x,y,z) = \frac{V_i}{V} \quad i = 1, 2, 3, 4$$

FIGURE 8.3   Geometric interpretation of the shape function of a tetrahedron element.

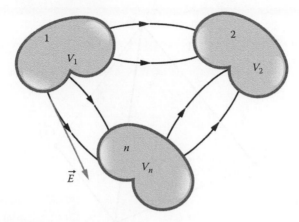

FIGURE 8.4 Typical eletrostatic problem.

8.3   3D ELECTROSTATICS

As mentioned in Chapter 4, the goal of a 3D electrostatics simulation is to determine the distribution of both the scalar electric potential function and the electric field vector generated by a set of excited conductors inside a dielectric medium that could be linear or not (Figure 8.4).

The starting point of this development is Maxwell's fourth equation—Gauss' law of electrostatics—given by

$$\oint_{\Sigma} \vec{D} \cdot d\vec{S} = Q_i \tag{8.7}$$

where
  $\vec{D}$ is the displacement vector (C/m^2)
  Q_i is the total amount of electrical charge inside the surface $\widetilde{\Sigma}$

The relation between the displacement vector and the electric field vector is the constitutive relation $\vec{D} = \varepsilon\vec{E}$, where ε is the electrical permittivity of the medium. The relation between both the electric field vector and the scalar electric potential is

$$\vec{E} = -\nabla V \tag{8.8}$$

8.4   3D INTERPOLATOR FUNCTION

The vertexes of the tetrahedron element in Figure 8.5 are numbered locally from (1) to (4), and we need to determine the potentials V_1, V_2, V_3, and V_4.

Using the same procedure we did in 2D electrostatic formalism, it is possible from (8.6) to represent the electric potential inside the element as follows:

$$V(x,y,z) = \sum_{i=1}^{4} N_i(x,y,z)V_i \tag{8.9}$$

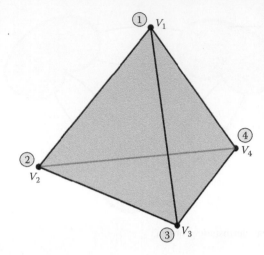

FIGURE 8.5 Generic element.

As $\vec{E} = -\nabla V$, the components of the electric field vector are given by

$$E_x = -\frac{\partial V}{\partial x} = -\frac{1}{6V}\left(b_1V_1 + b_2V_2 + b_3V_3 + b_4V_4\right)$$

$$E_y = -\frac{\partial V}{\partial y} = -\frac{1}{6V}\left(c_1V_1 + c_2V_2 + c_3V_3 + c_4V_4\right) \tag{8.10}$$

$$E_z = -\frac{\partial V}{\partial z} = -\frac{1}{6V}\left(d_1V_1 + d_2V_2 + d_3V_3 + d_4V_4\right)$$

With this approach, the electric field vector becomes constant inside the element, as the expressions in (8.10) show.

The next step in this formalism consists in involving each node of the 3D mesh by a polyhedron, the edges of which are composed of straight segments that link the element centroid to the middle of the edges.

Figure 8.6 shows in detail a generic element extracted from a 3D finite element mesh. The shaded region showed the portion of the polyhedron that involves node (1) of the element

Some faces of the polyhedron involving node (1) of element (e) are the triangular faces (M_2M_3G), (M_3M_4G), (M_4M_2G), where M_2, M_3, and M_4 are the points located at the middle of the edges (1,2), (1,3), and (1,4), respectively, and G is the centroid of the tetrahedron.

Then, for evaluating the flux of the displacement vector over the external surface of this polyhedron, we start evaluating the flux of the displacement vector over each portion of the surfaces of each element that admits node (i) as a vertex. That is,

$$\oint_\Sigma \vec{D} \cdot d\vec{S} = \sum \int_{S_i} \vec{D} \cdot d\vec{S} \tag{8.11}$$

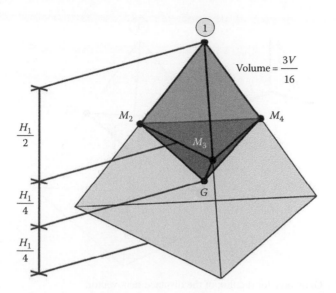

FIGURE 8.6   Generic finite element—some geometric dimensions are highlighted.

In this equation, S_i is the portion of the polyhedron surface that involves node (i). From Figure 8.6, we can see that the external portion surface S_i of the polyhedron is the one composed of the points $M_2 M_3 M_4 G$.

$$Q = \sum Q_i \qquad (8.12)$$

Here, the electric charge Q_i is the total amount of the electric charge in the shaded area in Figure 8.6. The summations from (8.11) and (8.12) have no null terms in the elements that admit node (i) as a vertex.

We can now evaluate the flux of the displacement vector over the surface $M_2 M_3 M_4 G$.

As the electric field vector is constant inside the element, the flux of the displacement vector in this faceted surface is the same flux over the triangular surface $M_2 M_3 M_4$ in Figure 8.7.

The flux of the displacement vector over the triangular surface $M_2 M_3 M_4$ is

$$\int_{Si} \vec{D} \cdot d\vec{S} = -D_x \Delta Syz - D_y \Delta Sxz - D_z \Delta Sxy \qquad (8.13)$$

The surfaces ΔSyz, ΔSxz, and ΔSxy are the projections of $M_2 M_3 M_4$ over the planes yz, xz, and xy, respectively.

From (8.10), it is possible to evaluate E_x, E_y, and E_z as follows:

$$D_x = \varepsilon E_x; \quad D_y = \varepsilon E_y; \quad \text{and} \quad D_z = \varepsilon E_z$$

We can demonstrate this in analytic geometry as follows:

$$\Delta Syz = \frac{b_1}{8}, \quad \Delta Sxz = \frac{c_1}{8}, \quad \text{and} \quad \Delta Sxy = \frac{d_1}{8}$$

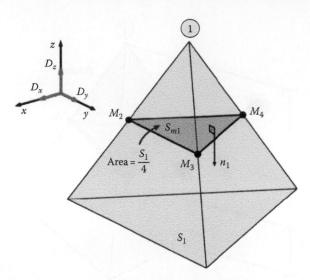

FIGURE 8.7   Geometry for the flux of the displacement vector.

Replacing these quantities in Equation 8.18 gives

$$\int_{Si} \vec{D} \cdot d\vec{S} = \frac{\varepsilon}{48V} \Big[\big(b_1 \cdot b_1 + c_1 \cdot c_1 + d_1 \cdot d_1 \big) V_1 + \big(b_1 \cdot b_2 + c_1 \cdot c_2 + d_1 \cdot d_2 \big) V_2$$

$$+ \big(b_1 \cdot b_3 + c_1 \cdot c_3 + d_1 \cdot d_3 \big) V_3 + \big(b_1 \cdot b_4 + c_1 \cdot c_4 + d_1 \cdot d_4 \big) V_4 \Big] \quad (8.14)$$

The same procedure should be applied in the evaluation of the flux of the displacement vector over other similar portions of the surfaces of the polyhedrons that involve nodes 2, 3, and 4.

The following matrix equation expresses the results obtained from the surface integration of the displacement vector, over the four portions of the closed surface involving each node of the mesh inside the generic element:

$$\begin{bmatrix} \displaystyle\int_{S_1} \vec{D} \cdot d\vec{S} \\[2em] \displaystyle\int_{S_2} \vec{D} \cdot d\vec{S} \\[2em] \displaystyle\int_{S_3} \vec{D} \cdot d\vec{S} \\[2em] \displaystyle\int_{S_4} \vec{D} \cdot d\vec{S} \end{bmatrix} = \big[C \big]_{4\times4} \cdot \big[V \big]_{4\times1} \quad (8.15)$$

where

$$C_{ij} = \frac{\varepsilon}{48V}\left(b_i \cdot b_j + c_i \cdot c_j + d_i \cdot d_j\right) \quad i,j = 1, 2, 3, 4$$

$$\left[V\right] = \left[V_1 \; V_2 \; V_3 \; V_4\right]^T$$

Assuming that the volumetric density of electrical charge is constant inside the element, the total amount of electric charge inside the volume with vertexes $1M_2M_3M_4$ is given by

$$Q_1 = \rho_v \cdot \text{Vol}(1M_2M_3M_4)$$

It is possible to demonstrate that

$$\text{Vol}(1M_2M_3M_4) = \frac{3}{16}V$$

where V is the volume of the tetrahedron given by

$$V = \frac{1}{6}\det \begin{vmatrix} 1 & x_1 & y_1 & z_1 \\ 1 & x_2 & y_2 & z_2 \\ 1 & x_3 & y_3 & z_3 \\ 1 & x_4 & y_4 & z_4 \end{vmatrix}$$

Therefore,

$$Q_1 = \frac{3V}{16}\rho_v$$

As the centroid with the other two vertexes and the middle edge's point of the element generate the same volume evaluated for the one that involves node (1), we can write the total amount of electrical charges inside the concerned volumes in the form

$$\begin{bmatrix} Q_1 \\ Q_2 \\ Q_3 \\ Q_4 \end{bmatrix} = \frac{3V}{16}\rho_v \begin{bmatrix} 1 \\ 1 \\ 1 \\ 1 \end{bmatrix}_{1\times 4} \tag{8.16}$$

We now have all the information needed for evaluating the flux of the displacement vector over the polyhedron's surface that involves each node of the domain using the following expression:

$$
\begin{bmatrix}
\oint_{\Sigma_1} \vec{D} \cdot d\vec{S} \\
\oint_{\Sigma_2} \vec{D} \cdot d\vec{S} \\
\cdot \\
\cdot \\
\cdot \\
\oint_{\Sigma_{NN}} \vec{D} \cdot d\vec{S}
\end{bmatrix}
=
\begin{bmatrix}
\sum \int_{S_1} \vec{D} \cdot d\vec{S} \\
\sum \int_{S_2} \vec{D} \cdot d\vec{S} \\
\cdot \\
\cdot \\
\cdot \\
\sum \int_{S_{NN}} \vec{D} \cdot d\vec{S}
\end{bmatrix}
=
\begin{bmatrix}
Q_1 \\
Q_2 \\
\cdot \\
\cdot \\
; \\
Q_{NN}
\end{bmatrix}
\tag{8.17}
$$

which, expressed in matrix form gives

$$
[C]_{NN \times NN} [V]_{NN \times 1} = [Q]_{NN \times 1}
\tag{8.18}
$$

To generate both the global matrix $[C]_{NN \times NN}$ and the actions vector $[Q]_{NN \times 1}$, we should apply the same algorithm used in Chapter 4.

As the equation's system (8.18) has the determinant null as explained in previous chapters, we need to introduce the boundary conditions in the same way discussed earlier.

After the introduction of the boundaries conditions, the equation becomes

$$
[\bar{C}]_{NN \times NN} [V]_{NN \times 1} = [\bar{Q}]_{NN \times 1}
\tag{8.19}
$$

The solution of (8.19) gives the electric potential of all nodes (vertexes of the tetrahedrons) of the mesh that enables us to evaluate the electric field vector inside each element. With this big mass of data, it is possible to estimate with good accuracy the performance of the device.

8.5   3D ELECTROKINETICS

3D FEM for electrokinetics is easy to understand if you are familiar with 3D electrostatics because they are analogous. The mathematical formulation for electrokinetics is obtained by comparing the equations for the two kinds of phenomena given in Table 8.1.

TABLE 8.1

Similarity between Electrostatics and Electrokinetics

Electrostatics	Electrokinetics
$\oint_{\Sigma} \vec{D} \cdot d\vec{S} = Q_i$	$\oint_{\Sigma} \vec{J} \cdot d\vec{S} = 0$
$\vec{E} = -\nabla V$	$\vec{E} = -\nabla V$
$\vec{D} = \varepsilon \vec{E}$	$\vec{J} = \sigma \vec{E}$
$\vec{D} \, (C/m^2)$	$\vec{J} \, (A/m^2)$
$\varepsilon \, (F/m)$	$\sigma \, (S/m)$

By making a simple substitution ($\vec{D}$ by $\vec{J}$ and ε by σ and establishing that $Q = 0$), Equation 8.18 can be written in the following matrix form:

$$\begin{bmatrix} \displaystyle\int_{S_1} \vec{J} \cdot d\vec{S} \\[2ex] \displaystyle\int_{S_2} \vec{J} \cdot d\vec{S} \\[2ex] \displaystyle\int_{S_3} \vec{J} \cdot d\vec{S} \\[2ex] \displaystyle\int_{S_4} \vec{J} \cdot d\vec{S} \end{bmatrix} = \left[G \right]_{4 \times 4} \cdot \left[V \right]_{4 \times 1} \tag{8.20}$$

where

$$G_{ij} = \frac{\sigma}{48V} \left(b_i \cdot b_j + c_i \cdot c_j + d_i \cdot d_j \right) \quad i, j = 1, 2, 3, 4$$

$$\left[V \right] = \left[V_1 \; V_2 \; V_3 \; V_4 \right]^T$$

The actions vector is null in this state (DC electric current) because we have no displacement current density, that is, $\vec{J}_d = \partial \vec{D} / \partial t = 0$.

With the introduction of the Dirichlet boundary conditions using the same procedure presented in Chapter 4, the final system's equation can be expressed as follows:

$$\left[\vec{G} \right]_{NN \times NN} \left[V \right]_{NN \times 1} = \left[\vec{I} \right]_{NN \times 1} \tag{8.21}$$

The solution of the system (8.21) gives the electric potential in all nodes (vertexes of the tetrahedrons) of the mesh that enables us to evaluate both the electric field vector and the current density vector inside each element. With this big mass of data, it is now possible to estimate with good accuracy the distribution of current in all 3D domains.

9 Results Exploration

9.1 INTRODUCTION

In the previous chapters, we presented the mathematical formulation of the finite element method (FEM) for electromagnetics derived directly from Maxwell's equations, which is an approach that allows both students and professionals who face it for the first time to understand it easier than the classical way proposed by other books, mainly because it does not involve any complex mathematical formalisms.

We aim with this book to diffuse this methodology because this is a powerful tool for the analysis and synthesis of electromagnetic devices.

The understanding of new information should always begin with an analysis of the elementary cases. With this, it is possible to acquire enough sensibility to move on to the next step and achieve complete knowledge of the technique. However, the simplicity presents limitations that the reader will identify as he/she gains more sensibility when using the technique.

This book only introduces the first-order FEM. This approach enables us to easily integrate Maxwell's equations inside an element. There is more evidence that the first-order element is less accurate [14,21,22,23,33,38] than the higher-order element (second order or more); thus, the first-order element become competitive with the progress of the computational machines that enable us to solve system's equation of hundred thousand or even more than million equations, which are big enough for good accuracy.

The core of FEM's package is its "solver" that corresponds not only to assemble the global system, but also to solve it. There are various algorithms for this task; however, we suggest using the incomplete Choleski conjugate gradients (ICCG) [10,13,24,25,29,35,52], because it is suitable to deal with sparse matrix, or the Bi-CCG (biconjugate gradient) [19,31,49,56,61,63,64] for complex problems.

9.2 RESULTS EXPLORATION

As the "solver" completes its task, the user will have a huge amount of data to analyze. The final results of the module are the coordinates of all nodes, information about all elements including the values of the potentials (electric or magnetic) in all vertexes. Given this big amount of numerical data, it is very difficult to extract any useful information.

Therefore, an additional module was conceived for translating numbers into images, graphs, and postprocessing computation routines for evaluating specific quantities depending on the study.

In 2D computation, it is very common to plot an equipotential line chart. This is important in analyzing whether the solution is correct.

Figure 9.1 shows an equipotential line chart of the magnetic field distribution in a cross section of an induction motor.

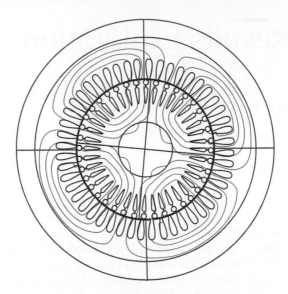

FIGURE 9.1   Flux lines in a magnetic field—4-poles induction motor.

FIGURE 9.2   Color shade chart—Each color represents a range of magnetic flux density field.

Figure 9.2 shows a color shade chart of the same motor. Here, the colors of each element of the cross section indicate the field (electric or magnetic) intensities in a range from soft to hard. This output is very important in identifying the regions where the limits supported by the material are reached [39].

Figure 9.3 shows two images. The first one is a 3D color shade chart of the magnetic flux density in a disk electric motor. The second one shows graphically the variations in the (z) component of the magnetic flux density in the air gap of the motor.

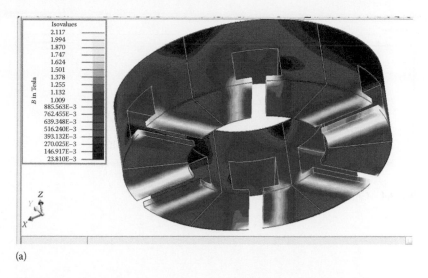

(a)

(b)

FIGURE 9.3 Features of an exploration module: (a) Color shade magnetic flux density. (b) Variation of the (z) magnetic flux density component.

As a postprocessing, the exploration module of an electromagnetic FEM package enables us to evaluate several quantities like parameters (R, L, and C) [57], magnetic losses, forces, torque, and others. Figure 9.4 shows the torque characteristics of the same disc motor in Figure 9.3.

The transient analysis exploration of electromagnetic phenomena enables us to evaluate the time evolution of both electric and magnetic flux densities at each time step and the time variations of the associated quantities [68].

Figure 9.5 shows the color shade magnetic flux density distribution in an open boundary device. We now see three images showing that distribution in three time instances for evidencing the features of this kind of simulation.

Finally, Figure 9.6 shows a 3D color shade in the nucleus of a three-phase power transformer.

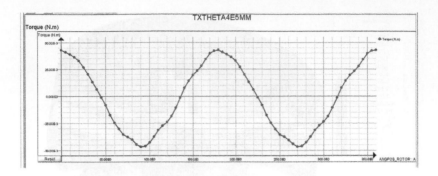

FIGURE 9.4 Characteristics of torque x angle.

(a)

(b)

(c)

FIGURE 9.5 Time variation of the magnetic flux density distribution: (a) $t = 0.5$ s.
(b) $t = 0.6$ s. (c) $t = 0.75$ s.

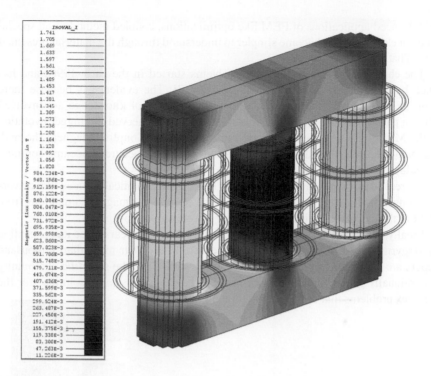

FIGURE 9.6   Flux density color shade of a three-phase transformer—3D domain.

9.3   CONCLUSION

The evolution of FEM in electrical engineering was completely different from that in both mechanical and civil engineering. In these fields, the method is introduced directly from the governing equations phenomena in that the physical interpretation is clear.

In electrical engineering, the introduction of FEM was based on both variational formulation, as proposed by M.V.K. Chari [4.67] and J.C. Sabonnadiere [35], and weight residual method, proposed by the others. The weight residual method [28] rapidly became the most important approach, because it is suitable for almost all kinds of phenomena over all time-dependent electromagnetic fields.

Maybe that is the reason the FEM is more familiar to mechanical and civil engineers [34,51,69] than to electrical engineers; if you are able to see the physical sense of a numerical method, it is very easy to understand. When we need to apply a complex mathematical formalism for demonstrating any numerical method, the physical sense is lost, and it becomes more difficult to learn it.

This book expects to demystify the notion that FEM for electromagnetics is based on complex mathematical formulations and uses simple techniques suitable for undergraduate students and beginners.

Advanced applications of FEM like optimizations, coupled methods, and special kinds of elements have become simpler to understand through the material presented here. That is our hope!

The challenge of solving coupled problems started in the early 1990s and has since become an important research line [25,37]. The evidence is because there is no an isolated phenomenon in the nature. All kinds of knowledge are involved when a physical event occurs. That is the reason we observed the bigger evolution of the Multiphysics Software packages in recent years. Coupling magneto-thermal, magneto-mechanical, and other science fields with high accuracy became common nowadays [37,59].

A great advancement in the simulation of electromechanical devices like drivers and converters for electrical machines was the coupling of magnetic equations from FEM to the electric circuit equations from drivers and/or converters.

Finally, it was not easy to arrive at its current status 60 years since its conception by Argyris, but FEM is still in its adolescence. Every year more than a thousand papers are published on the subject around the world, and the scientific community is continually growing. We are sure that FEM will give us more solutions for the complex problems that humanity is facing in this century.

References

1. Turner, M. J., Clough, R. W., Martin, H. C., Topp, L. C., Stiffness and deflection analysis of complex structures, *J. Aeronaut. Sci.*, 23(9), 805–823, 1956.
2. Zienkiewicz, O. C., *Finite Element Method in Engineering Science*, McGraw-Hill, London, U.K., 1971.
3. Kao, T. K., Hardisty, H., Wallace, F. J., An energy balance approach to finite element method applied to heat transfer analysis, *Int. J. Mech. Eng. Educ.*, 11(1), 1–19, 1983.
4. Silvester, P. P., Chari, M. V. K., Finite element solution of saturable magnetic field problems, *IEEE Trans. Power Syst.*, PAS-89, 1642–1651, 1970.
5. Silvester, P. P., Ferrari, R. L., *Finite Elements for Electrical Engineers*, 2nd edn., Cambridge University Press, Cambridge, U.K., 1990.
6. Hoole, S. R. H., *Computer-Aided Analysis and Design of Electromagnetic Devices*, Elsevier, New York, 1989.
7. Abe, N. M., Simulação de estruturas ferromagnéticas com imãs permanentes pelo método dos elementos finitos bidimensionais: Uma nova abordagem (in Portuguese), Master thesis, Escola Politecnica da USP, 1992.
8. Cardoso, J. R., A Maxwell's second equation approach to the finite element method applied to magnetic field determination, *Int. J. Elect. Eng. Educ.*, 24(3), 259–272, June 1987.
9. Jordão, R. G., A contribution to the analysis of magnetic circuits excited by permanente-magnets, *IEEE Trans. Educ.*, 38(3), 1995.
10. Pereira, F. H., Filho, B. A. R., Silva, V. C., Nabeta, S. I., Solution of nonlinear magnetic field problems by Krylov-subspace methods with *h*-cycle wavelet-based algebraic multigrid preconditioning, *IEEE Trans. Magn.*, 44(6), 950–953, 2008.
11. Lebensztajn, L., Silva, V. C., Rossi, L. N., Cardoso, J. R., Teaching electromagnetic fields and FEM for undergraduate students, *IEEE Trans. Educ.*, 44(2), 5–10, 2001.
12. Rossi, L. N., Cardoso, J. R., Silva, V. C., Lebensztajn, L., Silva, J. A. P., FEA of electromagnetic: A geometrical approach for problems with plane symmetry, *IEEE Trans. Magn.*, 38(2), 1313–1316, March 2002.
13. Chari, M. V. K., Salon, S. J., *Numerical Methods Is Electromagnetism.* Academic Press, San Diego, CA, 765pp., 2000.
14. Rothwell, E. J., Cloud, M. J., *Electromagnetics*, 2nd edn., CRC Press, Boca Raton, FL, 2009.
15. Balanis, C., *Advanced Engineering Electromagnetics*, Wiley, New York, 1989.
16. Feynman, R., Leighton, R., Sands, M., *The Feynman Lectures on Physics*, vol. 2, Addison-Wesley, Reading, MA, 1964.
17. Cheng, D. K., *Field and Wave Electromagnetics*, Addison-Wesley, Reading, MA, 1983.
18. Bastos, J. P., Sadowski, N., *Magnetic Materials and 3D Finite Element Modeling*, CRC Press, Boca Raton, FL, 2014.
19. Bastos, J. P., Sadowski, N., *Electromagnetic Modeling by Finite Element Methods*, Marcel Dekker, New York, 2004.
20. Van Bladel, J., *Electromagnetic Fields*, Hemisphere Pub. Co., New York, 1985.
21. Jackson, J., *Classical Electrodynamics*, John Wiley & Sons, New York, 1962.
22. Stratton, J. A., *Electromagnetic Theory*, McGraw-Hill Book Co., New York, 1941.
23. Al-Khafaji, A., Tooley, J., *Numerical Methods in Engineering Practice*, Holt, Rinehart and Winston, New York, 1986.

24. Sadiku, M. N. O., *Numerical Techniques in Electromagnetics*, CRC Press, Boca Raton, FL, 1992.
25. Jin, J., *The Finite Element Method in Electromagnetics*, John Wiley & Sons, New York, 1993.
26. Desai, C. S., Kundu, T., *Introductory Finite Element Method*, CRC Press, Boca Raton, FL, 2001.
27. Desai, C. S., Abel, J. F., *Introduction to the Finite Element Method*, Van Nostrand Reinhold, New York, 1972.
28. Finlayson, B. A., *The Method of Weighted Residuals and Variational Principles*, Academic Press, New York, 1972.
29. Logan, D. L., *A First Course in the Finite Element Method*, 5th edn., Cengage Learning, Stamford, CT, 2012.
30. Clough, R. W., Rashid. T., Finite element analysis of axisymmetric solids, *J. Eng. Mech. Div., Proc. Am. Soc. Civ. Eng.*, 91, 71–85, February 1965.
31. Cook, R. D., Malkus, D. S., Plesha, M. E., Witt, R. J., *Concepts and Applications of Finite Element Analysis*, 4th edn., Wiley, New York, 2002.
32. Pars, L., *An Introduction to the Calculus of Variations*, John Wiley & Sons, New York, 1962.
33. Bucalem, M. L., Bathe, K. J., *The Mechanics of Solids and Structures—Hierarchical Modeling and the Finite Element Solutions*, Springer-Verlag, Heidelberg, Germany, 2011.
34. Bathe, K. J., *Finite Elements Procedure*, Prentice Hall, Inc., Englewood Cliffs, NJ, 1996.
35. Sabbonnadiere, J. C., Coulomb, J. L., *Elements Finis et CAO*, Edition Hermes, Paris, France, 1986.
36. Cardoso, J. R., *Engenharia Eletromagnetica* (in porthuguese), Elsevier, São Paulo, Brazil, 2001.
37. Abe, N. M., Cardoso, J. R., Foggia, A., Coupling electric circuit and 2D FEM model with Dommel's approach for transient analysis. *Proceedings of 11th Conference on the Computation of Electromagnetic Fields-COMPUMAG/Rio*, Rio de Janeiro, Brazil, pp. 115–156.
38. Zienkiewicz, O. C., Taylor, R. L., *The Finite Element Method*, vol. 2, 4th edn., McGraw-Hill Book Company, London, U.K.
39. Claycomb, J. R., *Applied Electromagnetics Using Quick Field and MATLAB*, Jones and Barlett Publishers, Boston, MA, 2008.
40. Ulaby, F. T., *Electromagnetics for Engineers*, Pearson Education Inc., Upper Saddle River, NJ, 2005.
41. Krauss, J. D., *Electromagnetics*, 4th edn., McGraw-Hill, New York, 1992.
42. Ramo, S., Whinnery, J. R., Van Duzer, T., *Fields and Waves in Communication Electronics*, 3rd edn., John Wiley & Sons, New York, 1994.
43. Rao, N. N., *Elements of Engineering Electromagnetics*, 2nd edn., Prentice Hall, Englewood Cliffs, NJ, 1987.
44. Shen, L. C., Kong, J. A., *Applied Electromagnetism*, 2nd edn., PWS Engineering Books, Boston, MA, 1987.
45. Bansal, R., *Handbook of Engineering Electromagnetics*, Marcel Dekker, New York, 2004.
46. Inan, U. S., Inan, A. S., *Engineering Electromagnetics*, Addison Wesley Longman, Inc., Menlo Park, CA, 1999.
47. Miner, G. F., *Lines and Electromagnetic Fields for Engineers*, Oxford University Press, Oxford, U.K., 1996.
48. Franco, M. A. R., Passaro, A., Cardoso, J. R., Machado, J. M., Finite element analysis of anisotropic optical waveguide with arbitrary index profile, *IEEE Trans. Magn.*, 35, 1546–1549, 1999.

49. Andersen, O. W., Transformer leakage flux program based on finite element method. *IEEE Trans. Power Syst.*, PAS-92, 682–689, 1973.
50. Carpenter, C. J., Finite element network models and their application to eddy-current problems. *IEE Proc.*, 122, 355–362, 1975.
51. Zienkiewicz, O. C., Cheung, Y. K., Finite elements in the solution of field problems, *The Engineer*, September 1965, pp. 507–510.
52. Demerdash, N. A., Nehl, T. W., Flexibility and economics of the finite element and difference techniques in nonlinear magnetic fields of power devices, *IEEE Trans. Magn.*, MAG-12, 1036–1038, 1976.
53. Konrad, A., Silvester, P., Triangular finite elements for the generalized Bessel equation of order m, *Int. J. Num. Methods Eng.*, 7, 43–55, 1973.
54. Silvester, P., Construction of triangular finite element universal matrices, *Int. J. Num. Methods Eng.*, 12, 237–244, 1978.
55. Simkin, J., Trowbridge, C. W., On the use of the total scalar potential in the numerical solution of field problems in electromagnetics, *Int. J. Num. Methods Eng.*, 14, 423–440.
56. Norrie, D. H., Vries, G., *Introduction to Finite Element Analysis*, Academic Press, New York, 1978.
57. Simkin, J., Trowbridge, C. W., Three-dimensional non-linear electromagnetic field computations, using scalar potentials, *IEE Proc.*, 127, part B, 368–374, 1980.
58. Adams, R. A. M., *Sobolev Spaces*, Academic Press, New York, 1975.
59. Ahn, C. H., Lee, S. S., Hyuek, J., Lee, S. Y., A self-organizing neural network approach for automatic mesh generation, *IEEE Trans. Magn.*, MAG-27(5), 4201–4204, 1991.
60. AIbanese, R., Rubinacci, G., Solution of three dimensional eddy current problems by integral and differential methods, *IEEE Trans. Magn.*, MAG-24(1), 98–101, 1988.
61. Bathe, K. J., Wilson, E., *Numerical Methods in Finite Element Analysis*, McGraw-Hill, New York, 1980.
62. Manteuffel, T. A., An incomplete factorization technique for positive definite linear systems, *Math. Comput.*, 34, 473– 497, 1980.
63. Irons, B., Ahmad, S., *Techniques of Finite Element*, Ellis Horwood, Chichester, U.K., 1979.
64. Beckman, F. S., The solution of linear equations by the conjugate gradiente method, *In Mathematical Methods for Digital Computers*, vol. 1, John Wiley, New York, pp. 62–72, 1960.
65. Abe, N. M., Acoplamento Circuito Elétrico—Elementos Finitos em Regime Transitório Utilizando a Metodologia de Dommel (in Portuguese), PhD thesis, Escola Politecnica da USP, Sâo Paulo, Brazil, 1997.
66. Franco, M. A. R., Passaro, A., Sircilli Neto, F., Cardoso, J. R., Machado, J. M., Modal analysis of diffused anisotropic channel waveguides by a scalar finite elemento method, *IEEE Trans. Magn.*, 34(5), 2783–2786, 1998.
67. Silvester, P. P., Chari, M. V. K., Analysis of turbo-alternator magnetic fields by finite element, *IEEE Trans. Power Syst.*, PAS-90, 454–464, 1971.
68. Hannalla, A. Y., Macdonald, D. C., Numerical analysis of transient field problems in electrical machines, *Proc. IEE*, 123(9), 893–898, 1976.
69. Desai, C. S., *Elementary Finite Element*, Prentice Hall, New York, 1979.
70. Abe, N. M., Cardoso, J. R., Clabunde, D. R. F., Passaro, A., LMAG-2D: A software package to teach FEA concepts. *IEEE. Trans. Magn.*, 33(2), 1986–1989.

Index

Printed and bound by CPI Group (UK) Ltd, Croydon, CR0 4YY
01/11/2024
01782619-0020